OXFORD BIOLOGY PRIMERS

Discover more in the series at
www.oxfordtextbooks.co.uk/obp

Published in partnership with the Royal Society of Biology

GENOMICS

≋ OXFORD BIOLOGY PRIMERS

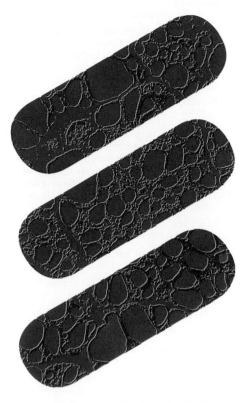

GENOMICS

Lia Chappell, Sarah J Lindsay, Phil Jones, Julian Parkhill, Jonathon Roberts,
Nancy Holroyd (with Faye Rodgers), Michal Szpak
(with Yali Xue and Chris Tyler-Smith), and Francesca Gale

Edited by Ann Fullick
Editorial board: Ian Harvey, Gill Hickman, and Sue Howarth

OXFORD
UNIVERSITY PRESS

Great Clarendon Street, Oxford, OX2 6DP,
United Kingdom

Oxford University Press is a department of the University of Oxford.
It furthers the University's objective of excellence in research, scholarship,
and education by publishing worldwide. Oxford is a registered trade mark of
Oxford University Press in the UK and in certain other countries

Published in the United States of America by Oxford University Press
198 Madison Avenue, New York, NY 10016, United States of America

British Library Cataloguing in Publication Data

Data available

Library of Congress Control Number: 2019954545

ISBN 978-0-19-884838-7

Printed in Great Britain by
Bell & Bain Ltd., Glasgow

PREFACE

Welcome to the Oxford Biology Primers

There has never been a more exciting time to be a biologist. Not only do we understand more about the biological world than ever before, but we're using that understanding in ever more creative and valuable ways.

Our understanding of the way our genes work is being used to explore new ways to treat disease; our understanding of ecosystems is being used to explore more effective ways to protect the diversity of life on Earth; our understanding of plant science is being used to explore more sustainable ways to feed a growing human population.

The repeated use of the word 'explore' here is no accident. The study of biology is, at heart, an exploration. We have written the Oxford Biology Primers to encourage you to explore biology for yourself—to find out more about what scientists at the cutting edge of the subject are researching, and the biological problems they're trying to solve.

Throughout the series, we use a range of features to help you see topics from different perspectives.

Scientific approach panels help you understand a little more about 'how we know what we know'—that is, the research that has been carried out to reveal our current understanding of the science described in the text, and the methods and approaches scientists have used when carrying out that research.

Case studies explore how a particular concept is relevant to our everyday life, or provide an intimate picture of one aspect of the science described.

The bigger picture panels help you think about some of the issues and challenges associated with the topic under discussion—for example, ethical considerations, or wider impacts on society.

More than anything, however, we hope this series will reveal to you, its readers, that biology is awe-inspiring, both in its variety and its intricacy, and will drive you forward to explore the subject further for yourself.

ABOUT THE AUTHORS

Lia Chappell, BA (Hons), MSci (Hons), PhD, chapter 7

Lia Chappell is a molecular biologist who loves to tackle a tricky problem. She currently works on the biochemical reactions, lab equipment, and analysis methods that are used for measuring active genes in single cells, using technology that allows thousands of cells to be processed in parallel. Prior to that, much of her work has been focused on getting sequencing-based assays of active genes to work well for the highly AT-rich genomes of malaria parasites. Lia has a degree in biochemistry, and is also actively involved with science outreach in Cambridge, and was the student president of Cambridge Hands-On Science (CHaOS) during her PhD. She has also tried a range of other outreach activities, including communicating science by the medium of stand-up comedy!

Francesca Gale, BSc (Hons), MSc, chapters 1, 2, and 7

Francesca grew up living in the Suffolk countryside. From the age of four, when given her first animal encyclopaedia, she was fascinated by animals. This passion for the natural world led Francesca to study biological science at the University of Exeter. After university, she joined Colchester Zoo, running the Education Department. This job brought together all things she enjoyed—travel, working with animals, and talking to different audiences about biodiversity, wildlife, and conservation. After eight years, she studied part-time for a Masters degree in science communication at the University of the West of England. This opened the door to a whole new field of science communication, and she joined the Wellcome Trust Sanger Institute as Education Officer. She is now Education Development Lead for the Wellcome Genome Campus Connecting Science, working alongside scientists to develop engaging resources for school students of all ages, and leading professional development courses on genomics for science teachers.

Nancy Holroyd, BSc (Hons), chapter 5

Nancy Holroyd grew up with a love of wildlife and biology thanks to her Dad, who often took her out to traipse across muddy fields when she was a child. She studied zoology at Imperial College, London, where her favourite subjects by far were molecular biology, genetics, and parasitology. Before, during, and after her bachelor's degree, Nancy worked at the then Sanger Centre sequencing the yeast genome, and later piecing together the genome assemblies of many pathogenic organisms. After several years working in different teams in the Sanger Institute, Nancy had the opportunity to combine her genomics experience with her fascination with parasites, working on numerous projects centred around the genomes of parasitic worms, particularly worm species responsible for a number of Neglected Tropical Diseases (NTDs). The research into understanding the genomes of parasitic worms will underpin much future research into tackling NTDs and breaking the poverty cycle in many countries across the world. In turn, Nancy now traipses her family across muddy fields whenever possible!

Phil Jones, BA, BM BCh, PhD, FRCP, FMedSci, chapter 2

Phil trained in medicine and then specialized in cancer treatment. He took time out to do a PhD on stem cells in the skin. Since then he has done lab research and clinical medicine treating cancer patients. Currently, Phil leads a research group at the Wellcome Sanger Institute and is Professor of Cancer Development at the University of Cambridge. Phil's group studies how mutations in single stem cells in normal tissues may give rise to cancer in order to discover better ways to prevent the disease. He also sees patients with skin cancers at Addenbrooke's Hospital in Cambridge.

Sarah J Lindsay, BA (Hons) MSc PhD, chapter 1

Sarah wanted to be an archaeologist when she was at school, and focused on arts subjects for GCSE and A level, going on to study for a degree in archaeology and prehistory at the University of Sheffield. After graduation, Sarah heard that the newly founded Sanger Centre (now the Wellcome Sanger Institute) was looking for technical staff to work on human and nematode genome projects, which sounded interesting and completely different to anything she'd done before. It was a particularly exciting time in genomics; the science at the institute was constantly changing, there was so much to learn, and the scientists were inspirational. After eighteen months at the institute, she recognized it was time to get some formal qualifications in science in order to advance her career. She left the institute to take a Masters degree in biomolecular archaeology, followed by a PhD in molecular biology at the University of Manchester. After this, Sarah returned to the institute to work in sequencing research and development, and then in human genetics, where she has contributed to our understanding of genetic variation and *de novo* mutations that occur in the genomes of mice and humans.

Prof Julian Parkhill, BSc (Hons) PhD FMedSci FRS, chapter 4

Julian Parkhill was interested in science from an early age, but was turned on to biology by an especially engaging teacher during O- and A-levels (thank you, Mr Hanks!). He went to Birmingham to study biology, intent on a career in research, much to the disappointment of his other teachers, who thought he should study medicine. For his PhD in Bristol, he investigated bacterial genetic switches controlling resistance to mercury. After working for a while on human viruses and cancer, he was lucky to land a job at the Sanger Centre in the very early days of bacterial genomics, contributing to the completion and analysis of the first bacterial genome sequenced in the UK (*Mycobacterium tuberculosis*). He stayed at the Sanger Centre, sequencing many more bacterial pathogens. He adapted new technologies as they were developed to study large populations of pathogens, trying to understand the emergence, transmission, and evolution of pathogenic and drug-resistant bacteria. He has collaborated with many others to use this technology for basic science as well as to drive it into the clinic for patient benefit. In 2019, he moved to the University of Cambridge to continue this research and to help teach the next generation of scientists.

Dr Jonathon Roberts, BSc (Hons), MSc (Hons), PhD, chapter 3

Jonathan started his career with a degree in the history and philosophy of science from University College, London. He is now an NHS genetic counsellor and also an academic researcher at the Wellcome genome campus (Society and Ethics Research). His current research explores ethical issues about genetics. His works also looks at science communication and genetics and he is particularly interested in the representation of genetics, inheritance, and DNA in popular culture. Jonathan's recent research explores how families' knowledge and enjoyment of pop culture can be used to facilitate engagement with genomics.

Jonathan works as a clinical genetic counsellor in the NHS. His clinical experience includes providing genetic counselling in cancer, cardiac, prenatal, and ophthalmology clinics. In addition to his genetic counselling practice, he has also worked as a teacher in special needs schools (in both the UK and Poland) and as a volunteer for the Samaritans.

Faye Rodgers, MA, MRes, PhD, chapter 5

Faye did her undergraduate degree in natural sciences at the University of Cambridge, specializing in genetics. She then moved to Imperial College London for an MRes (Master of Research), working mostly on bacterial pathogens, followed by a PhD on the genetics of malaria vector mosquitoes. Since finishing her PhD, she has been working as a bioinformatician at the Wellcome Sanger Institute, where she is a member of the Parasite Genomics team. She works mostly on the WormBase ParaSite website, which stores, analyses, and presents helminth genomic data to the research community.

Michal Szpak, BSc (Hons), MSc (Hons), PhD, chapter 6

Following an early childhood dinosaur obsession, Michal developed a fascination for evolutionary history and the amazing diversity of life. He took a degree in biology at the University of Warsaw, where he studied ancient DNA of extinct Pleistocene mammals. He subsequently moved to bioinformatics and explored human genetic variation across different ethnicities in his master's degree research at the Center for Public Health Genomics, University of Virginia. He eventually joined the Human Evolution team at the Wellcome Sanger Institute in 2012 and received his PhD in human population genetics from the University of Cambridge. His doctoral and postdoctoral work at the Sanger focused on positive selection and genetic adaptations in humans. He investigates the link between genetic variants and diverse adaptive traits found in human populations around the world. He also developed a passion for public engagement and has opportunities to communicate science to a wider public through arts, exhibitions, and lectures.

Chris Tyler-Smith, PhD, chapter 6

Chris grew up on a small farm in south-west England, where his lifelong interest in the natural world around him and its history began. After a degree in biochemistry at the University of Oxford and a PhD in molecular biology at the University of Edinburgh, he began a career in scientific research. This started by studying the DNA sequences needed to transmit human chromosomes when cells divide, but soon moved to using DNA variation to investigate the history and diversity of humans around the world. In 2003, he established the Human Evolution group at the Wellcome Sanger Institute near Cambridge, where this work expanded from small-scale bench experiments studying single genes to today's big science, where massive teams collaborate to sequence thousands of ape or human whole genomes.

Yali Xue, PhD, chapter 6

Dr Yali Xue studied public health as an undergraduate, epidemiology for her Masters degree, and medical and population genetics for her PhD, all in Harbin Medical University, China. While she was there, she worked on Chinese human genetic diversity, collecting samples from different ethnic groups in China and establishing cell lines from them, many of which are now included in the major worldwide panel used to investigate human genetic diversity. She joined the Human Evolution group at the Wellcome Sanger Institute in 2004, where she has worked on using variation on the Y chromosome to provide insights into human history and evolution, and patterns of variation throughout the genome to detect signatures of positive selection and derive evolutionary insights from both modern and ancient DNA.

University of Cambridge, Department of Veterinary Medicine

The Department of Veterinary Medicine at Cambridge is at the forefront of veterinary science and education and is a centre of excellence for teaching and research. Its mission is to improve the prevention and treatment of diseases of animals through the best clinical practice, by understanding and developing the science underpinning best practice, and by embedding an education programme in the veterinary sciences that delivers the best veterinary practitioners, academics, and research scientists.

Wellcome Genome Campus Connecting Science

Connecting Science is based at the Wellcome Genome Campus, Hinxton, Cambridgeshire. Comprised of four teams: Public Engagement, Advanced Courses and Scientific Conferences, Society and Ethics Research, and the Wellcome Genome Campus Conference Centre, it connects researchers, health professionals, and the wider public, creating opportunities and spaces to explore genomic science and its impact on people.

Wellcome Sanger Institute

The Wellcome Sanger Institute is one of the world's leading genome centres. Through its ability to conduct research at scale, it is able to engage in bold and long-term exploratory projects that are designed to influence and empower medical science globally. Institute research findings, generated through its own research programmes and through its leading role in international research collaborations, are being used to develop new diagnostics and treatments for human disease.

CONTENTS

ABBREVIATIONS

aDNA	ancient DNA
BES	bloodstream expression site
cDNA	complementary single-stranded DNA
dbSNP	database for single nucleotide polymorphisms
DDD study	Deciphering Developmental Disorders study
DECIPHER	DatabasE of generiC varIation and Phenotype in Humans using Ensembl Resources
DTC	direct-to-consumer genetic testing
EGFR	epidermal growth factor receptor
EHEC	enterohaemorrhagic *E. coli*
ExAC	Exome Aggregation Consortium
GM	genetically modified
gnomAD	Genome Aggretation Database
gRNA	guide RNA
GWEP	Guinea Worm Eradication Programme
HEG	homing endonuclease gene
HPO terms	human phenotype ontology terms
HUS	haemolytic-uraemic syndrome
IGF-I	insulin-like growth factor-I
kb	kilobase
LCA	last common ancestor
Mb	megabase
MDA	mass drug administration
mRNA	messenger RNA
MRSA	methicillin-resistant *Staphylococcus aureus*
NGS	next-generation sequencing
NTDs	neglected tropical diseases
PCR	polymerase chain reaction
QTL	quantitative trait locus
RNAi	RNA interference
SCBU	Special Care Baby Unit
SNP	single nucleotide polymorphism
SNV	single nucleotide variations
TALEN	transcription activator-like effector nuclease
UPEC	uropathogenic *E. coli*
VSG	variant surface glycoprotein
WES	whole-exome sequencing
WGS	whole-genome sequencing
WSS	Wiedemann-Steiner syndrome
ZFN	zinc-finger nuclease

1 RARE DISEASES: A GENOMICS PERSPECTIVE

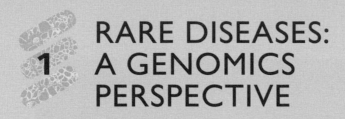

When you hear the term **genomics**, what comes to mind? It is a term that was almost unknown outside the scientific community twenty years ago. Now the word 'genomics' is widely used in the media and every school pupil learns what it means. However, if you ask someone on the street for a definition, you are still likely to get answers ranging from 'Is it to do with DNA or genes?' to 'Is it about garden ornaments?'. So, these questions remain: what is genomics and why study it?

Genomics in its broadest definition is the study of the structure and function of genomes, and a genome is the complete set of genetic instructions for an organism (see Fig 1.1). It is a huge area of study that encompasses many different (and sometimes controversial) scientific techniques, which together can help us understand ourselves and the world around us. Over the past century there have been many technological advances that now make it possible for us to delve deep into our own genomes and even to make changes to our own DNA and that of other organisms—many of these techniques are examined in chapter 7 of this book.

This genomic revolution is set to have a massive impact on human society, from understanding diseases, to improving health care, finding new drugs and therapies, and understanding our past and possibly our future. But genomics isn't just about humans: genomics can also help us understand the complex biodiversity of life on earth and how we as a species are having an impact on our planet. All of these topics are explored by experts in their field within this book. In this chapter, we look at the impact of genomics, starting with how they have changed the way we view some of the most unusual human diseases.

Fig. 1.1 Genomics—the study of the complete set of genetic instructions of all living organisms.

Natali_ Mis/Shutterstock.com

What is genomics?

Scientists have known for many years that DNA contains a mixture of four bases (adenine, cytosine, guanine, and thymine) within the backbone of a double helix structure, and they have also known the proportions of these four bases for the DNA of many species. So why is it useful to know the order of the nucleotides? What value does it have converting these bases into letters that can be encoded as text?

Knowing the order of the nucleotides enables us to decode the order of the amino acids in protein sequences—though identifying these protein sequences within the DNA isn't always straightforward. With some knowledge of the DNA sequences recognized and bound by proteins we can also identify sequences which don't encode proteins, but instead control when and where the synthesis of proteins is activated. Comparing DNA sequences between many individuals allows us to link trends for 'spelling variants' of DNA sequences with certain traits, such as height, hair colour, or the risk of developing a specific disease.

We can now use a range of technologies to capture and read DNA and related molecules from many kinds of cells, including technologies sensitive enough to measure DNA from individual cells. Price drops in the costs of such technology have led to DNA sequencing being applied in almost all areas of biology, and increasingly within health care too. Further advances in these sequencing technologies will enable us to monitor DNA in many different environments—in the future, it's likely that we will be able to watch forecasts of likely viral pathogens in our local area as routinely as it is now possible to check weather forecasts.

If we can read DNA, can we also edit or write DNA? We can, and these methods are constantly improving. Until recently, editing DNA sequences within a living cell was slow and laborious, but new technologies now make this much faster and easier. Improvements have also been made to the technology for making lengths of synthetic DNA within the laboratory. These developments are helping to advance the area of research known as synthetic biology, where researchers have already managed to make entirely artificial genomes of very simple organisms by stitching together lengths of overlapping synthetic DNA. These tools could make many additional technologies possible—for example, we may be able to produce bacteria that can synthesize drugs designed by researchers.

Deciphering the genome

The development of techniques that are able to decode the sequence of bases in a DNA molecule underlies many of the breakthroughs in genomics and disease research over the past few decades. These methods have enabled scientists to sequence the genomes of many different species, including humans, animals, plants, and microbes. Today, hundreds of thousands of people across the world have their genomes read with the aim of identifying the causes of a whole range of diseases.

Deciphering the genome—understanding the millions of bases and the sequences in which they are arranged—is an incredibly complex undertaking. We would not be where we are today without the endeavours of many talented and accomplished scientists; there is enough material for a whole book on the history of genomics alone. Here, we will highlight several key scientific milestones of the twentieth century which provided the foundations for the genomic revolution we are experiencing today.

- In 1953, James Watson and Francis Crick, with contributions from Rosalind Franklin and Maurice Wilkins, discovered the double helix structure of DNA, and subsequently, in 1958, Crick proposed the theory of the Central Dogma: the process by which the instructions in DNA are converted into a functional product.

- In the 1970s, Frederick Sanger (see Fig 1.2) developed the first DNA sequencing method, a process used to determine the order of DNA bases. Sanger and his team used this to sequence the first full genome of a virus called phiX174. This technological advance paved the way for the Human Genome Project.

- The Human Genome Project began in 1990. It was a global effort to sequence the first human genome, with research centres in the USA, UK, Germany, France, Japan, and China all working collaboratively together. It finished early and cost less than expected—but it still took thirteen years and cost around 1 billion US dollars.

The Human Genome Project has left a strong scientific legacy, and paved the way for many large, collaborative sequencing projects such as the

Fig. 1.2 Frederick Sanger won the Nobel prize for his work on sequencing DNA—work that has led directly or indirectly to all of the other developments described in this book.

Courtesy U.S. National Library of Medicine

1,000 genomes project, UK10K, and most recently the 100,000 genomes project. We have seen technological advances in DNA sequencing methods that have cut the time and cost of sequencing, making it feasible to sequence whole populations (see Fig 1.3). New sequencing methods mean that sequencing is no longer confined to the lab: portable mini-sequencers can be taken to the rainforest, Antarctica, and the International Space Station. We can even re-program, edit, and sequence different cell types to help us understand disease at a cellular level. You can explore some of these developments both here and in chapters 2, 4, 5, and 6.

The many technical advances have also led to medical advancements from understanding and tracking drug resistance in bacteria and parasites to the development of personalized medicines in the treatment of cancer and other diseases. In 2017, Dame Sally Davies, the UK's Chief Scientific Advisor at the time, produced a white paper outlining her vision for 'Generation Genome', where genomics is embedded into the National Health Service with, for example, cancer patients routinely being offered DNA analysis of their tumours to help select the best treatments for them.

This sounds like significant progress—and it is, when you consider that 100 years ago, we didn't even know the structure of DNA or its role in inheritance. But there is still a long way to go. And in the pursuit of using genomics for a happier, healthier, and more genome-aware society, there need to be checks and balances. Probably the most significant legacy of the Human Genome Project is the Ethical, Legal and Social Implication (ELSI) framework that now underpins all genomics research. It ensures privacy and fairness in the use of genetic information and considers the potential

Fig. 1.3 Iceland is one of the first countries in the world to take steps towards sequencing its entire population. Although there are only around 350,000 Icelanders, this is still a massive undertaking.

© Anthony Short

implications of the introduction of new genetic technologies, such as genetic testing, into the clinic. When dealing with any genomic data, we must consider how that data is used and accessed, to protect the privacy of the individuals who have contributed their DNA. More of these ethical and social implications of genome sequencing will be addressed in chapter 3. But here, in this chapter, we are going to focus on the impact genomics has had on our understanding of rare diseases.

What is a rare disease?

Rare diseases are just that: rare. They don't affect many people and they are notoriously hard to diagnose. When only a handful of people around the world suffer from a particular disease, it's not surprising doctors don't always get to the bottom of why a patient is suffering.

Yet rare diseases are a significant burden to the individuals affected and the society they live in, not least because some patients may experience chronic illness and disability and possibly premature death. Many of those affected are undiagnosed, leaving their families with a personal burden of worry about what the future may hold, and concern about whether future children may also be affected.

Unfortunately, it can be difficult to diagnose rare diseases correctly because many disorders look superficially similar (have similar phenotypes), and the way doctors interpret these phenotypes can vary. The disease mechanism itself can be complex, making diagnosis even harder because the same disease looks subtly different in different individuals. Sadly, without a diagnosis at the molecular or genetic level, many patients are never diagnosed at all. As a result, they remain untreated, without effective therapeutic care and not understanding why they are affected in the way that they are.

Perhaps this sounds a bit depressing—and it has been depressing for sufferers for thousands of years. But the study of genomics is beginning to shine a light on these rare diseases (see Fig 1.4): we can be more optimistic about the future!

Fig. 1.4 The instructions within the genome are followed through fetal development and after birth—a tiny error in the wrong place can lead to rare but catastrophic changes in the way the body functions.

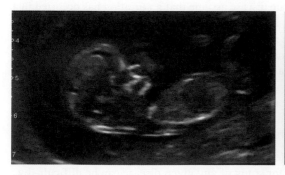

© Anthony Short

There are official definitions of what constitutes a rare disease. In the USA, a disease is considered rare if it affects fewer than 200,000 people at any given time. In Europe, a rare disease is one that affects fewer than 1 in 2,000 individuals. Despite these diseases being individually rare, they are collectively common; approximately 6–7 per cent of the UK population is affected by a rare disease—so this amounts to around 60 million people with a rare disease in the USA and the European Union alone. Rare diseases vary in prevalence: some may affect only a few individuals worldwide, whereas others are found in thousands of people. Sadly, children are disproportionately burdened: 75 per cent of rare diseases affect children. Even more sadly, the reason for this is that many rare diseases mean that the affected children never make it to adulthood.

At the time of writing, we have defined around 7,000 rare diseases, but only around 50 per cent of them have had the molecular cause determined. However, it is likely that many more rare diseases might actually exist but we simply haven't described them yet. Some of these diseases are shared by only a few individuals globally, so it is only in the modern era of electronic communication that links between these patient phenotypes are being made.

At the moment, our poor understanding of rare disease is hindering diagnoses and effective treatment. However, in the years since the human genome was published in 2001, the rate at which we have been discovering new disease genes has rapidly increased. This has been driven by a combination of new sequencing technologies such as **next-generation sequencing (NGS)** (see chapter 7), large research studies into rare disease, and the sharing of clinical and phenotypic data in the research community. Every new gene that is discovered helps us to better understand the molecular basis of disease, potentially leading to immediate health benefits for those diagnosed with rare diseases. It is an exciting and rapidly developing field; in the near future faster and more accurate diagnoses will be possible due to the comprehensive databases of clinical and phenotypic profiles that are being developed, and the expert bioinformatics support which will be enabling them. As the price of sequencing continues to fall, and the size and portability of sequencing technology improves, eventually genetic diagnoses should be available to everyone, everywhere.

Finding the causes of rare diseases—past and future

Our current models for rare disease assume that a **mutation** has occurred in a gene causing the disease. However, this simple concept is much more complicated than it appears. We need to be sure which parts of the genome are genes and we need to understand the function of the protein the gene codes for.

If this wasn't complicated enough, not all mutations in a single gene are the same; not all mutations affect the resultant protein. Therefore, in order for a mutation to be the cause of a rare disease, it needs to be very rare or co-occur and co-act with another **variant**.

There are other barriers to rare disease research:

- Collecting a sufficient number of patients to achieve statistically significant results for study of any given disease can be difficult—by definition, there are not many of them!

- Funding for research into very rare diseases can be hard to find. There is less interest in the therapeutic value for rare diseases, compared to common diseases, not least because there is much less profit to be made from treating a small number of patients.

- Population structure can have a confounding effect. Genetic variants may vary in prevalence in different populations for historical reasons. We know that if populations are isolated, are very closely related (**consanguineous**), have had historical migrations, or have experienced **population bottlenecks**, the frequency of rare disease alleles in the population can be affected.

A population bottleneck is an event or a series of events that dramatically reduces the size of a particular population, causing a large reduction in the size of the gene pool, and so in the genetic diversity of the population. This in turn often results in an increased incidence of particular alleles in the population going forward. These alleles are those which happened to be present in the population that survived the bottleneck (see Fig 1.5).

One example of this is Gaucher disease, which causes a number of varying symptoms in those affected, including anaemia and bone disease. It occurs in between 1:50,000 and 1:100,000 individuals in the general population, but in around one in twelve of those with Ashkenazi Jewish ancestry. One hypothesis is that this is due to a population bottleneck that occurred 1,000 years ago in the ancestors of the modern Ashkenazi community. The Ashkenazi community is very aware of these raised genetic risks, and many communities have screening programmes for young people planning to have a family, to discover whether they are carriers.

A number of studies have applied a targeted approach to exploit this kind of population structure to help us understand rare disease. The project 'East London Genes and Health' was set up to recruit individuals from local communities with a high level of parental relatedness. One of the aims is to discover knockout genes, which will help us to understand which genes

Fig. 1.5 High levels of Ellis van Creveld syndrome in the Amish people in the US (a) are the result of a population bottleneck in the eighteenth century (b).

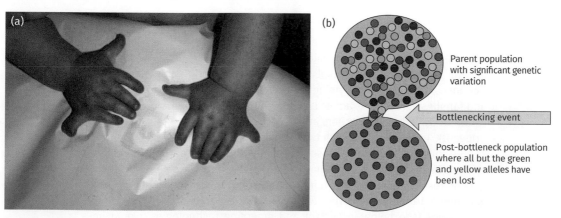

a) Reproduced from Geneviève Baujat and Martine Le Merrer, Ellis-Van Creveld syndrome. *Orphanet Journal of Rare Diseases,* 2007;2:27. © Baujat and Le Merrer; licensee BioMed Central Ltd. 2007. Attribution 2.0 Generic (CC BY 2.0) b) Tsaneda/ Wikimedia Commons

we can manage without. Some populations do not have access to the latest technologies and have not yet been well characterized genetically, making it potentially harder to find the causes of rare disease in those populations.

Until recently, finding the molecular cause of a rare disease was a labour- and resource-intensive process. **Linkage analysis** was the most commonly used technique; this process looks for parts of the genome that always segregate with the disease phenotype in a family of affected individuals. Linkage analysis can be done by assuming a model of inheritance for the disease (dominant or recessive) and tracking the inherited disease through a pedigree. If there is a large enough sample size, we don't need a clear model of inheritance: statistically significant evidence of allele sharing across affected individuals is enough to explain what is observed. If we know that a gene, or group of genes, have a function that is biologically relevant to the way the disease develops and the symptoms it causes, then those 'candidate' genes can be prioritized as part of this approach. This technique was used to isolate the genes for disorders such as cystic fibrosis and neurofibromatosis, but it could take many years to find a gene for a single syndrome or rare disease using this method.

However, the situation has improved hugely over the past ten years, due to technological advances and collaborations between clinicians and scientists. To find the mutations that cause disease, we need to have a good understanding of the current genetic variation in 'healthy' individuals. The availability and decreasing costs of NGS (see chapter 7) has enabled large sequencing projects to take place, establishing a background of common human variation. The first of these was the 1,000 Genomes project, which ran between 2008 and 2015, creating a large, open-source catalogue of human variation data across twenty-six populations. More recently, a number of studies have been set up with the specific aim of finding the causes of rare diseases. You can discover the difference a study like this can make in Case study 1.1.

Case study 1.1
DDD study and CDK13

Imagine how it feels if your child is born with a deformity, has abnormal growth patterns, or intellectual disability—and no one can tell you why. Genomics studies are beginning to answer these questions for scientists and families alike. One example is the Deciphering Developmental Disorders (DDD) study, led by researchers at the Wellcome Sanger Institute near Cambridge UK. In collaboration with twenty-four regional clinical genetic centres, 12,000 children with undiagnosed developmental disorders were recruited for the study. The children had already been through standard UK NHS genetic testing (which, at the time of the research, screened only for large genomic rearrangements). The researchers wanted to test if NGS and **array techniques** (see below in Current and emerging technologies) would improve diagnostic rates in individuals who had not already received a genetic diagnosis. The individuals recruited were referred with a severe developmental disorder and were normally the only affected family member.

The DDD study used arrays and **exome sequencing** to look for potentially causative mutations in patients—exomes are the protein-sequencing regions of genes. It combined this with systematic phenotyping of the patients using standardized **Human Phenotype Ontology terms (HPO terms)**. In addition, growth measurements, family history, and developmental milestones were recorded for each affected child.

The researchers found that, on average, each individual in the study had two **de novo** (new) mutations in a coding or splicing sequence. The researchers constructed a **null model** for the number of new mutations they would expect to see in each gene, based on our understanding of the rate and sequence context at which de novo mutations occur. They used data on the distribution of de novo mutations from previously generated data in unrelated studies (~3,000 individuals who had a range of disorders from intellectual disability to autism) to help them do this. They then tested to see which genes had more mutations than they would expect. They found ninety-three genes that had a statistically significant excess of de novo mutations, and eighty of these had been previously associated with a developmental disorder.

The researchers used a combination of **animal modelling** and risk scores to corroborate their theories that the mutations they observed caused the disease. Clinicians already knew that facial similarities are often evident in individuals with rare diseases (see Fig Aii). Therefore, to make it easier to share their findings, the researchers created anonymized composite faces

Fig. A Tamika and Caitlyn (Ai) both have a mutation of their CDK13 gene which occurred spontaneously as they developed. The girls have played a generous part in the Deciphering Developmental Disorders study coordinated at the Wellcome Trust Sanger Institute near Cambridge. Boys born with Barth syndrome (Aii) have consistent facial features with a tall, broad forehead, round face with full cheeks, prominent ears, and deep-set eyes. Scientists have observed that children with similar rare diseases often share similar facial features.

(Ai)

(Aii)
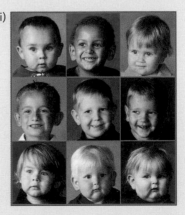

Reproduced from Clarke, Sarah L. N., et al., Barth syndrome. *Orphanet Journal of Rare Diseases*, 20138:23. © Clarke et al.; licensee BioMed Central Ltd. 2013. Attribution 2.0 Generic (CC BY 2.0)

from two-dimensional clinical photos of patients, along with visual summaries of the phenotypes associated with each disease gene. This data can be viewed by anyone in the open-access online resource DECIPHER.

The DDD study estimated that de novo mutations accounted for approximately one-half of severe developmental disorders in the study group and concluded that they arise in approximately one in every 300 births. This means that around 400,000 children are born with a rare disease globally every year. The timing of the de novo mutation is important; if it takes place in a specific ovum or sperm, the parents of an affected child will go on to have unaffected children if they decide to have more babies. However, sometimes an individual contains populations of cells with different genotypes as a result of mutation during their development. This individual may have mutations that are present in their germ cells (sperm or eggs) but not in their other tissues. This is called **mosaicism**. If one of the parents has mosaicism, there is a risk that more of their gametes may be affected, even though the parent is healthy. As a result, the disease may recur if the parents decide to have more children.

One of the genes that the team found to be associated with a new genetic disorder was CDK13. Only eleven children in the UK have been diagnosed with this disorder. The main symptoms seen in children with this disorder are facial abnormalities and developmental delay, but there are a number of different additional phenotypes ranging from abnormalities in hands and feet to seizures.

The study found that Tamika Kyd (aged ten) and Caitlyn Hunter (aged nine) had mutations in the CDK13 gene, causing their genetic disorder. The families live in London and met for the first time as a result of the study; they were surprised at how much the girls resembled each other (see Fig Ai). Despite the strong resemblance between the girls, they do not have exactly the same phenotype: Tamika has a heart defect while Caitlyn does not, and Tamika has more language and communication difficulties than Caitlyn.

This shows that, for families affected by rare disease, there can be great personal value to being able to share experiences with each other. In addition, the fact that the DDD study established that the mutations were de novo in Caitlyn and Tamika meant that the families could go ahead and have more children without the risk of recurrence.

❓ Pause for thought

Why is systematic phenotyping of patients so important?

Apart from the clinical value, can you think of other reasons for sharing data?

In 2013, the UK government set up 'Genomics England', a company owned and funded by the Department of Health and Social Care. The aim was to sequence the whole genomes of 100,000 individuals, some of whom have been diagnosed with one of a panel of 190 rare diseases. The pipeline currently prioritizes rare diseases with unmet diagnostic needs and individuals with extreme forms of common disease. The initial phase of the project uses trios (a family structure), but in the future, as we learn more and more about common variation, it may be possible to diagnose disease in a single individual without the need for family information. Aside from increasing diagnosis rates, the project also serves to develop an infrastructure for future use in the NHS, and of course there is the potential for lifelong electronic health record linkage with the data, enabling longitudinal studies.

So, we have established that around 50 per cent of rare diseases are caused by de novo mutations in genes. But what about the remaining cases? New computational methods are being developed and datasets assembled to answer this question, and they are discussed below.

Current and emerging technologies

Until recently, studies used a candidate gene approach, where a gene/genes of interest were identified and then interrogated using targeted sequencing. With the advent of cheap whole-genome technologies, the techniques became much less limited.

The first whole-genome studies used array technologies. An array is a chip containing a panel of known variants. An individual's DNA would be hybridized to the chip, so if the individual had one of the variants on the chip they would be identified. The studies would generally have a number of patients (cases) and screen them alongside a number of populations of matched 'healthy' individuals (controls). Although these can only capture common variation, they can capture genetic associations (ie not the disease-causing variant itself but a haplotype that may be associated with it), along with large genomic rearrangements that can cause rare disease.

These methods have now largely been replaced by whole-exome sequencing (WES) and whole-genome sequencing (WGS). Whole-exome sequencing involves capturing and sequencing the coding part of the genome. Only 1-2 per cent of the total genome actually codes for proteins, so this is currently the more cost-effective way of finding genetic variants that are likely to be associated with gene expression. Scientists can identify a number of different genetic variations this way, from single-nucleotide mutations, through to copies of whole chromosomes.

As the amount of sequencing data grows, *in silico* (computational) methods are being rapidly developed to help us understand the data better. These often provide a score, or rank, for a variant or gene, and often are developed using machine learning in conjunction with large datasets. For example, PolyPhen is a predictive tool that predicts the effect of a mutation on the structure and function of the protein. It uses information on known

7. Comments from parents of individuals diagnosed with Wiedemann-Steiner syndrome: http://wiedemannsteiner.com/about-wss/.

8. Genes & Health: http://www.genesandhealth.org/.

9. Germline gene therapy: https://www.yourgenome.org/debates/is-germline-gene-therapy-ethical.

10. Genome-wide association studies: https://www.yourgenome.org/stories/genome-wide-association-studies.

11. Gene therapy: https://www.yourgenome.org/stories/treating-the-bubble-babies-gene-therapy-in-use.

Discussion questions

1.1 Do you think the collection and sharing of data on patients with rare disease is skewed towards particular populations? If so, what might that mean?

1.2 Suggest reasons why a data breach in any of the databases mentioned in the study would be a serious issue.

1.3 Some people have suggested that rare diseases should be screened for during antenatal testing. Discuss the benefits and problems of introducing this.

1.4 What do you think 'good' and 'bad' uses of gene therapy could be? Consider reasons why other people might not agree with you.

2 CANCER GENOMICS

Cancer is a potentially fatal disease that impacts the lives of almost everyone, directly or indirectly. All of us will know someone who has had cancer, as one in three people develop the disease. As treatment improves, however, more and more cancer sufferers are being cured. Understanding of every aspect of cancer, from how it develops to which treatments are best for individual patients, is being revolutionized by DNA sequencing. Cancer genomics is one of the fastest-moving areas of medical research and it is having a direct impact on people's lives.

In adults, all of our tissues maintain a constant number of each cell type within them. To achieve this, cell division matches the rate at which cells are lost from the tissue. Cancer is a disease in which cells divide in excess, generating a lump, known as the primary tumour. A key feature of cancer is that the cells in the lump spread, invading the neighbouring normal tissues and blood and lymphatic vessels, allowing them to colonize distant organs, forming distant secondary tumours. For example, a cancer that starts in the bowel may spread to the liver and lungs. Both primary or secondary tumours may cause symptoms as they penetrate and grow into normal tissues (see Fig 2.1). These symptoms can include pressure on nerves causing pain, or disruption of the function of a tissue like the lungs or liver, causing breathlessness or jaundice as the burden of tumours increases. Here we consider how understanding the genome of cancer cells can help us prevent and treat cancer, and improve the survival and quality of life of patients.

Fig. 2.1 This diagram summarizes the main stages in the spread of a primary cancer. Uncontrolled cell growth generates a mass of cells called a tumour, which then starts to invade the surrounding tissues. The tumour cells eventually invade blood vessels, allowing them to spread to distant parts of the body and establish tumours in other organs, a process called metastasis.

Mutation causes cell to grow abnormally

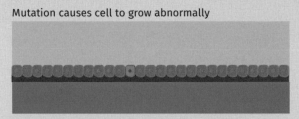

Uncontrolled cell growth leads to tumour formation

Growing tumour invades surrounding tissue

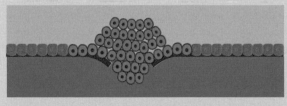

Cancer cells are able to travel in blood

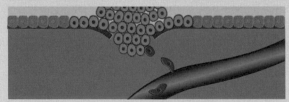

Secondary tumours form in other parts of the body

The genetic origins of cancer

Over the past few years there have been some major developments in genome sequencing, making it faster and cheaper—see chapter 7 for the details of these new technologies. These developments have had some far-reaching and perhaps unexpected benefits in the way we understand cancers and how they develop. Sequencing the genomes of a large number of cancers of different types has revealed some common features they all share.

First, all cancers have alterations in their genome. Childhood cancers may have only one or two changes, whereas in adult cancers, multiple changes occur. The types of changes are much like the errors you might make if you were asked to type out the 3,200,000,000 Cs, Gs, As, and Ts that make up the human genome. You would start by making single-letter errors, like typing a T when you should write a G. Eventually, as you got more sleep deprived you might miss out a whole section or type out the same section several times. You might even get so exhausted that you cut-and-pasted one section of the text into another where it didn't belong. The number of errors would increase if having finished doing it once, you then had to copy it out again, using the version you had just finished typing out.

The tired typist analogy helps explain how normal cells accumulate genetic alterations. A cell that divides once per month for eighty-five years over a human life span will have to copy its genome over 1,000 times. Each mistake passed on to the two daughter cells at division will be carried on through the subsequent generations of cells. There is very good evidence that many cancers come from a single normal cell that has picked up alterations in its genome over multiple generations, until one of its distant descendants accumulates sufficient changes to become a cancer (see Fig 2.2).

Fig. 2.2 A healthy lung (left) and a lung affected by cancer (right)—it is astonishing that tiny changes in the DNA code can have such devastating effects on the body.

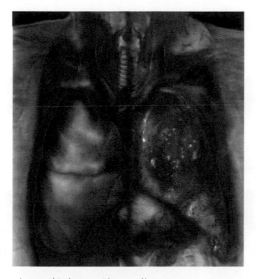

Anatomical Travelogue/Science Photo Library

Single nucleotide variations

The same types of genetic errors are found in almost all cancers. Following the typing model, a single letter mistake, such as switching a C for a T, is the most common error. These changes are termed mutations or **single nucleotide variations (SNVs)**. It has been estimated that in normal skin cells, a new mutation occurs every few days of life (see Fig 2.3). Most SNVs don't affect cell behaviour. This may be because the nucleotide change is in a region between genes or in a gene that is not expressed in the cell type that is mutated. These neutral mutations are termed 'passengers', carried along each time the cell divides but not increasing the risk that it or its daughters will transform into a cancer.

The SNVs that do matter are ones which occur in the parts of genes encoding proteins that alter cell behaviour in a way that increases the chance of subsequent cell generations transforming into tumour cells. These mutations are termed 'drivers' and they are found in a high proportion of cancers of a given type.

The most common driver SNVs in cancer affect a gene called *TP53*, which controls the production of a protein called tumour protein p53. *TP53* is central to the quality control of the genome, so if a mutation occurs in it, it's like turning off the spell checker. Each cell has a set of proteins that detect damaged DNA. If this is detected, TP53 protein (sometimes also referred to as p53) is activated, which either halts cell division until the damage is fixed or triggers the cell to self-destruct in a process known as apoptosis. Without functional TP53, cells are more likely to transform into cancer, as they can accumulate mutations and other genetic changes without hindrance.

TP53 is just one of many of genes that defend cells against cancer. Genes like this are called **tumour suppressors**. They regulate processes like cell division, controlling checkpoints through which a cell has to pass in order to divide. Cells have multiple levels of such safety mechanisms that have to

Fig. 2.3 This diagram models a gene with an SNV, a single-letter 'typo', which can result in cancer.

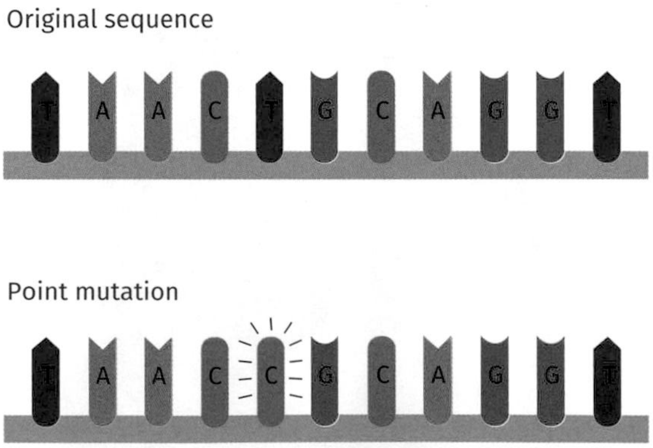

Original sequence

Point mutation

be turned off in order to develop cancer. Tumour suppressor genes are very commonly found to be inactivated in cancer genomes. For most genes this requires the loss of both copies (alleles) of the gene. *TP53* is unusual, as just a single mutation can disrupt its function. This is because it acts as a complex of four protein molecules that work together. A mutation of one allele may result in a change in shape of TP53 protein, so any complex containing one or more mutant proteins will not work normally. The vulnerability of working as a tetramer, along with its critical role in maintaining quality control of the genome, helps to explain why *TP53* is so commonly mutated in cancers.

Other driver mutations alter the activity of proteins that regulate other processes, such as cell proliferation. A very commonly mutated gene in some cancers is *KRAS*, which links signals from growth factor receptors at the cell surface to events in the cell nucleus that trigger entry into the cell cycle. KRAS protein is usually switched off, unless turned on by signals from growth factors. The mutations are clustered around the twelfth and thirteenth amino acids of the KRAS protein, and result in the protein being permanently switched on, driving the mutant cells to divide. Other common mutations alter the activity of growth factors themselves, leaving proteins like the epidermal growth factor receptor (EGFR) in a permanently active state.

Deletions

As well as making single-letter mistakes when typing, it's also possible to highlight and delete a large chunk of text, as most of us know from bitter experience. Deletions are a common feature of cancer genomes. These may be of a few nucleotides or may extend to part or all of a chromosome containing many genes. If one of the genes that is lost is a tumour suppressor, mutation or loss of the other allele will increase the chance of a cell developing cancer.

Amplifications and whole genome duplication

As well as cutting out text, you can highlight pages and paste them in multiple times. When this occurs in the genome, it's called amplification (see Fig 2.4). Amplifications can be small, perhaps only of a single gene, but more commonly they affect dozens or hundreds of genes. One example of a frequently amplified gene is the *EGFR*. This occurs commonly in adult brain cancer, resulting in an excess of *EGFR* signals that drive the cancer cells to divide.

Another abnormality found in cancers is whole-genome duplication, where the number of genomes in the cell rises from the normal two to four. The impact of this change is still unclear.

Translocations

A final example of an alteration in a cancer genome that can be replicated by a typist is a cut-and-paste error: randomly dropping part of a page to a whole chapter of text into a place where it doesn't belong. Most

Another example of a carcinogenic chemical is found in some Chinese traditional herbal medicine. It is called aristocholic acid and it causes a distinctive change from A to T in DNA. Sequencing of liver and bladder tumours from Taiwan and China shows a substantial proportion contain a high number of mutations due to long-term herbal medicine use. It is hoped that banning the use of plants containing aristocholic acid will cut the incidence of these cancers.

Fig. 2.5 People are not the only animals who can develop cancers in response to viral infections. For example, chickens are prone to develop several cancers linked to specific viruses, and research into these mechanisms has been useful for both animal and human health.

© Anthony Short

Fig. 2.6 Mutational signatures—such as the changes seen in skin cells exposed to ultraviolet radiation - can change the sequence of key cancer-related genes to make new proteins, for example switching the normal *TP53* gene to make a protein with a tryptophan (W) instead of arginine (R) at the critical amino acid 248. This may result in skin cancers like the melanoma shown here (b). Reading mutational signatures helps scientists identify some of the causes of cancer. Almost all skin cancer DNA changes are due to UV light, so it makes sense that using sun protection will cut the risk of the disease.

(a)

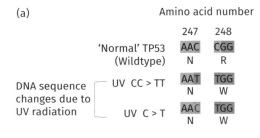

	Amino acid number	
	247	248
'Normal' TP53 (Wildtype)	AAC N	CGG R
UV CC > TT	AAT N	TGG W
UV C > T	AAC N	TGG W

DNA sequence changes due to UV radiation

(b)

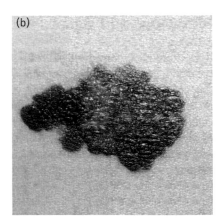

Key to amino acids
N = asparagine
W = tryptophan
R = arginine

National Cancer Institute

While mutational signatures show us how we can avoid some cancers—for example, by avoiding exposure to the sun and using sun block on our skin when we go outside, by avoiding smoking, and by not using dubious herbal remedies—other signatures are due to processes within cells that may damage our DNA, and these we have no control over. These cell-intrinsic signatures will generate mutations even if we avoid exposing ourselves to things that mutate DNA, though such 'healthy living' will cut our risk of getting cancer.

Family cancers

Early in cancer research it was realized that in some families many people develop cancer of a particular type. By sequencing the DNA of affected and unaffected family members it was possible to show that this was sometimes due to some family members inheriting a mutated copy of a tumour suppressor gene. This means that if the other copy of the gene is deleted or mutated, there is an increased risk of developing cancer.

The first gene identified in this way was the **retinoblastoma** gene, *RB*. This encodes a protein that regulates cell division. Retinoblastoma is a very rare tumour of the back of the eye that develops in young children. Somewhat amazingly, it has been diagnosed from photos at children's parties. Usually both eyes will show 'red eye', which comes from flashlight reflected from the retina. In retinoblastoma, one eye is red and the other white, because of the tumour on the retina. The tumours develop because one defective allele of *RB* was inherited from a parent and the other has become inactivated by mutation in the eye.

Since *RB*, many other tumour suppressors have been found by studying cancers that run in families. For example, the risk of breast and ovary cancer is strongly increased in some families who carry mutations in the breast cancer-linked genes *BRCA1* and *BRCA2*, genes that are involved in repairing damaged DNA. People who inherit a defective copy of *TP53* have an increased risk of multiple cancers. Often tumour suppressors found in family cancers are also mutated or otherwise lost in 'sporadic' cancers of the same tissues. As genome sequencing becomes faster and cheaper, our ability to identify these familial cancers and, at the very least, offer susceptible people additional screening, is growing all the time.

How cancer develops

Scientists think that many cancers develop over a long period from mutated cells in normal-looking tissues. Once a single cell has picked up a mutation, it will be passed on to its daughter cells and so on potentially for many generations, forming a 'clone', a group of cells that all come from the original mutated cell. Recently, it has become clear that normal human skin in sun-exposed areas of the body is a patchwork of such clones, each of which carries one or more mutations linked with skin cancer (see Fig 2.7). Some of these clones are very large, extending for over a centimetre. The clones in normal human skin carry the ultraviolet signature, showing

Fig. 2.7 Ageing normal human skin is a patchwork of cells containing different mutations linked to cancer (drivers), shown here by coloured circles; open circles carry passenger mutations that are not related to cancer.

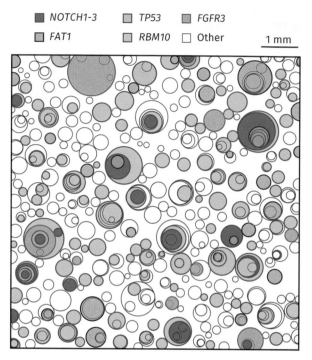

■ *NOTCH1-3* ■ *TP53* ■ *FGFR3*
■ *FAT1* ■ *RBM10* □ Other

1 mm

Reproduced with permission from Martincorena, I et al, High burden and pervasive positive selection of somatic mutations in normal human skin. *Science.* 2015 May 22;348(6237):880-6. Copyright © 2015, American Association for the Advancement of Science. DOI: 10.1126/science.aaa6806.

that the mutation they share was caused by sunlight. Intriguingly, only a few mutated genes generate large clones, and these seem to have been selected by a process of natural selection similar to that which operates in the Darwinian evolution of new species of animals or plants.

How do we know this? The evidence comes from the mutations. In the evolution of a species, a gene that is being selected for will have many mutations that alter the structure and function of the protein it codes for, compared with those mutations that are silent (or synonymous) and so do not change the protein. This is also the case for mutant clones in normal skin, suggesting that the large clones are the evolutionary winners in a process of natural selection by making the cell 'fitter' than the surrounding normal cells. The specific protein changes caused by the selected mutations are very similar to those seen in cancers, consistent with cancer developing from these clones. In someone of fifty, at least one-third of cells in their normal skin have cancer-associated mutations.

Evidence from cancer genome sequencing as well as experiments on cancer formation in cell cultures suggest that multiple genome changes are needed to generate almost all cancers, so cancer remains predominantly a disease of old age. These changes may accumulate by one of the cells in a clone carrying, for example, a *TP53* mutation picking up a second genome change. If that change improves the fitness of the cell, it may persist and spread through the tissue. This process may be repeated until enough mutations and genome alterations have accumulated to form an early cancer, one that is confined to its normal tissue. A further DNA change may trigger invasion and metastasis, a full-blown cancer.

Sequencing thousands of cancers has revealed that of all the 20,000 genes in the human genome, only a few are frequently found mutated in cancers, and under normal Darwinian selection. Some, such as *TP53*, are mutated in the majority of all cancers. Others are more restricted to particular types of cancer. For example, a gene called *PTCH1* is mutated in almost all cases of a common kind of skin cancer called basal cell carcinoma, while another gene, *BRAF*, is mutated in a different kind of skin cancer called melanoma. Instead of just classifying cancers by the tissue they come from and the type of normal cell they resemble, it is increasingly possible to group them by the mutations they carry. This is of profound importance for treatment, as we discuss now.

Why don't we all have cancer?

The question posed in this heading might sound odd. One-third of us will get cancer at some point in our lives, though an increasing proportion of us will survive it. But we all have a trillion or more cells, many of which divide hundreds of times over a lifetime. If cancer-linked mutations and other genome changes give cells a competitive advantage, they may spread widely through a normal tissue, generating a large number of cells that have taken a first step towards cancer. Viewed against the large numbers of mutated cells, it seems surprising that more people don't get cancer.

There are multiple reasons for this. First, the fact that we have two copies of each gene means that for the vast majority of genes, both have to be mutated or deleted for the gene to be lost. Second, our cells have evolved multiple layers of defence to protect their genome. These include proteins that detect different types of DNA damage and organize its repair. For example, in the skin, ultraviolet light striking DNA may generate two types of chemical alteration, known as cyclobutane pyrimidine dimers (CPD) and 6–4 photoproducts (PP). If not repaired, these changes can block the expression of a gene by preventing its transcription into RNA. Proteins scan the DNA for the presence of abnormalities like cyclobutene pyrimidine dimers and 6–4 photoproducts, detecting them because they alter the shape of the DNA molecule. A complicated process then occurs, involving cutting out the abnormal nucleotides and filling in the resulting gap using the other DNA strand to ensure the repaired DNA sequence remains correct by pairing As with Ts and Cs with Gs.

The whole process is called nucleotide excision repair (see Fig 2.8). Its importance is illustrated by people who inherit mutations in one of the key proteins involved. They suffer numerous skin cancers if they go out in the sun and have to avoid sunlight—they may even end up only going out at night in the dark to avoid further cancers developing.

Fig. 2.8 If these repair mechanisms do not work properly, we are at a relatively high risk of developing cancer.

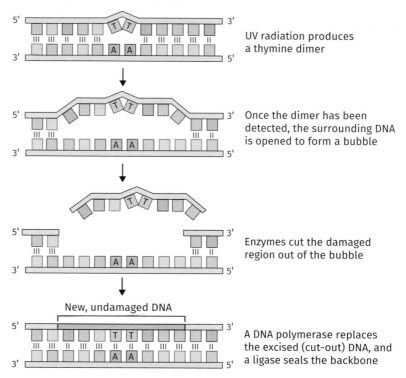

UV radiation produces a thymine dimer

Once the dimer has been detected, the surrounding DNA is opened to form a bubble

Enzymes cut the damaged region out of the bubble

New, undamaged DNA

A DNA polymerase replaces the excised (cut-out) DNA, and a ligase seals the backbone

Other kinds of DNA damage can also be repaired. The most dangerous DNA damage is a 'double-strand break', which may be caused by X rays or certain chemicals but may also happen as the cell copies its genome prior to division. Both DNA strands are broken, making it more likely that a cell may develop a translocation. A dedicated set of proteins is used to fix double-strand breaks. This may be done by joining the broken ends together, a process that involves the breast cancer genes *BRCA1* and *BRCA2*. Sometimes the cell is able to make a perfect repair of a double-strand DNA break, but it is possible to introduce errors in the genome in the repair process. Such 'error-prone' repair is frequent in cancers, where it may increase the number of gene alterations in cancer cells.

Another line of defence against cancer are 'checkpoints', like road blocks a cell can only pass through if its genome is intact. These are set up at key steps in the cell division cycle, as you can see in Fig 2.9. If a cell has detected DNA damage, its journey towards division will be halted until repairs are complete. Mutations that disable the checkpoints and allow genome damaged cells to divide increase the chances of developing cancer.

Cells also have an inbuilt self-destruct mechanism, called programmed cell death or apoptosis (see Fig 2.10). If triggered, an enzyme cascade destroys the cell's organelles and nucleus in an orderly way. Unrepaired DNA damage and other stresses, including those caused by some cancer-linked genomic changes, can trigger apoptosis, deleting the at-risk cell.

Fig. 2.9 Every cell passes through the cell cycle at different rates, controlled at a number of checkpoints. If these checkpoints fail, control over cell division can be lost, resulting in the development of cancer.

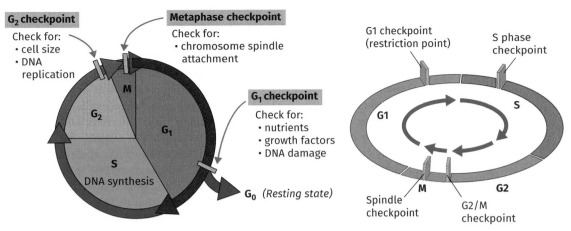

Adapted with permission from Craig, N et al, *Molecular Biology* 2e, Oxford University Press. Reproduced with permission of the Licensor through PLSclear

Fig. 2.10 Apoptosis, or programmed cell death, plays a vital role in removing cells with damaging mutations that might lead to cancers.

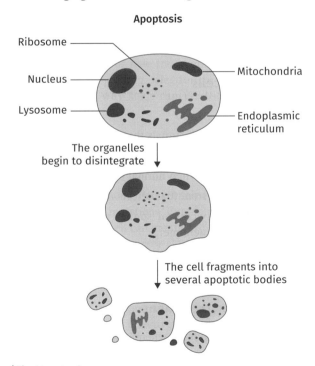

In order to develop into a cancer, cells typically need four or more genome alterations that disable cell safety systems. Only when these have been turned off is cancer development made possible.

treatment. Like chemotherapy, the effective targeted inhibitors also have side effects. Vismodegib causes painful muscle cramps and makes food tasteless, while vemurafenib can give skin rashes, and whilst effective against melanoma, may cause the growth of another kind of skin cancer called squamous cell carcinoma.

Fig. 2.12 Knowledge of the genomic basis of some cancers is enabling us to develop highly targeted drugs, such as an inhibitor of a common mutation BRAFV600E in melanoma skin cancers.

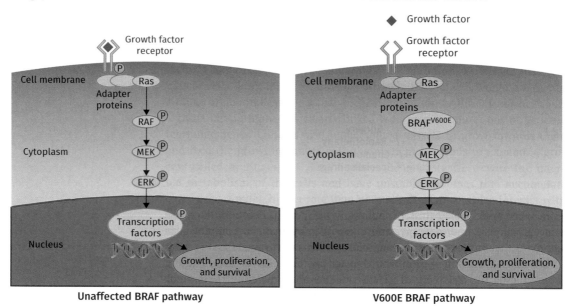

Using genomics to guide diagnosis

Working out which type of cancer a patient has is sometimes very difficult, but is crucial if the best treatment is to be selected. Only a small piece of cancer tissue may be available, and the type of cancer may not be clear by looking at the cells inside it. In cases where making a diagnosis is difficult, it is becoming common to sequence one or more genes. Examples include the presence of specific translocations creating fusion genes like *BCR–ABL* in leukaemia or relocating genes like *BCL2* in lymphoma. Childhood cancers may be very difficult to diagnose, as several types contain small cells that stain blue. The presence of particular translocations can help define the type and the treatment. An example is Ewing's sarcoma, where most cases contain a translocation that creates a fusion of two genes, *EWS* and *FLI1*. These both encode transcription factors, proteins that alter the transcription of many other genes, and the fusion protein alters levels of specific target genes to drive normal cells into Ewing's sarcoma. The *EWS–FLI1* fusion gene can be detected in clinical samples and used to confirm the diagnosis.

Using genomics to plan treatment

In many kinds of cancer, the treatment that is given is guided by scans that show whether the cancer has spread and what the cells look like when a piece of the cancer is looked at down the microscope by a pathologist. Based on large clinical trials, a treatment is selected. This process is rather hit and miss. Some patients who would have done well anyway may receive too much treatment that does not benefit them, whereas others are treated but don't respond. Sequencing genes in the tumour can help tailor the treatment to the individual.

Examples include sequencing the *BRAF* gene in melanoma. Only patients who carry an activating *BRAF* mutation will benefit from vemurafenib treatment; those without require alternative treatments. Similarly, activating *EGFR* mutations occur in about 10 per cent of cases of lung cancer. Patients whose cancers carry mutant *EGFR* may gain substantial benefit from specific *EGFR*-inhibiting drugs, but they are completely ineffective in those who do not have the specific mutations targeted by the drug. In breast cancer, about 20 per cent of patients have an amplification of the *HER2* gene, which codes for a protein receptor related to *EGFR*. This is important, because those patients with *HER2*-amplified breast cancers may be treated with specific antibodies against *HER2* such as trastuzumab (see Fig 2.13). Testing for mutations to guide treatment choices is now routine in several cancers and it is becoming more common in rarer cases to look for genome alterations that might be sensitive to treatment (see Case study 2.1).

Following responses to treatment

A key challenge in treating cancer is to try and get rid of each and every cancer cell, so that the disease doesn't simply regrow. It is very useful to have highly sensitive ways to detect a small number of cancer cells that may have survived treatment. This may be done by testing for genetic changes specific for the cancer, such as the specific translocations mentioned earlier in leukaemia cells. Following treatment, a sample of the patient's bone marrow is extracted—this is where any remaining leukaemia cells will be found. The DNA from the cells in the bone marrow is analysed by highly sensitive methods to detect a very small proportion of altered genomic DNA if it is present, as this would indicate that some cancer cells have survived.

More recently, it has been found that many cancers release small amounts of DNA into the blood stream. These may be detected in a blood sample and used to track the response of a cancer to treatment. The effectiveness of methods to look at this circulating tumour DNA is a subject of very active research, but the benefits of such tests have yet to be proven in most cancers.

Fig. 2.13 The use of gene alterations to diagnose cancer is increasing rapidly, and has the potential to enable doctors to begin the most effective treatment as rapidly as possible.

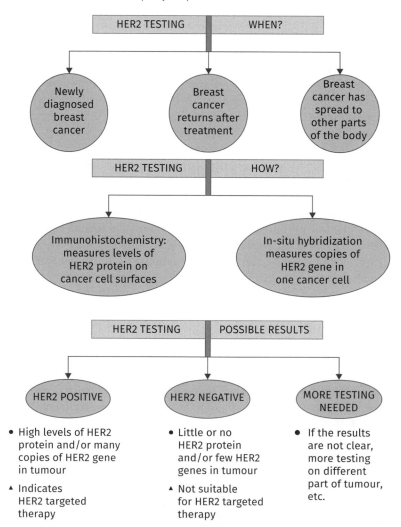

Why treatment fails

The sad truth about treating cancer is that the treatment may fail. Genomics has started to uncover why this happens. A cancer originates from the daughters of a single cell that have picked up multiple genome alterations. However, from this 'trunk', different branches may grow out, each of which has additional DNA changes, occupying different parts of the cancer. The trunk commonly has mutation or deletions of tumour suppressor genes like *TP53*, while the branches may have different activating mutations of growth factor receptors like *EGFR* or of genes in their downstream signalling pathways. It follows that a treatment that kills one or two of the branches of the tree may not deal with the others that have a different mutation, and they will survive to spread and mutate even further (see Fig 2.14).

The problem of new genetic changes conferring the ability to survive is exacerbated by the unstable genome found in many cancers which generates new mutations and genome alterations. As a result, the trunk has many branches, some of which may thrive, while others die back during treatment. Any mutation that enables cells to survive a drug will give a survival advantage and be selected for during treatment. This can be seen as an evolutionary process, the survival of the fittest clones under the selective pressure of a drug. The populations of the resistant cells expand and when after a short time the disease comes back, the patient will no longer respond to the drug (see Fig 2.14).

Understanding cancer genomes has revolutionized our understanding of the basis of the disease and has led to the development of new cancer treatments. Reading mutational signatures will help to lower the chances of

Fig. 2.14 Cancer isn't a single disease—even within one patient, the cancer cells mutate and change. The upper portion of the figure shows a tumour in which all cells share driver mutations (blue) that caused the tumour to develop, but some parts of the tumour have developed additional mutations. The development of the mutations may be represented as an evolutionary tree. The lower portion shows the effect of a treatment that kills all the cells except those that have mutation 3, which enables them to survive. Over time these resistant cells will regrow and a different treatment will be needed to kill them.

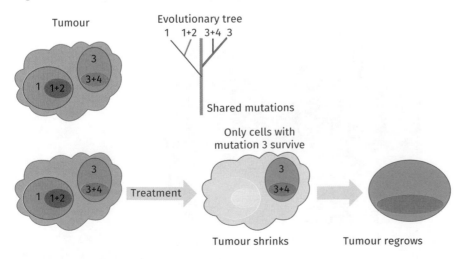

people getting cancer by highlighting avoidable cancer risks. The treatment of cancer will be improved both by the discovery of drugs to treat specific genetic changes and by identifying which cancers will respond to existing cancer treatments. Cancer genomics is thus one of the most exciting areas of genetic research in which discoveries are making an immediate impact on human health.

Case study 2.1
Cancer genomics and the discovery of new treatments for cancer

Several recent advances have made it possible to move beyond using genome sequencing to just describe the mutations in cancers, allowing us to take on the challenge of how to treat cancers with particular mutations most effectively. If the combination of driver genome alterations in each tumour can be defined by sequencing, can the most effective treatment for that combination be identified so the patient can receive the most effective treatment?

To take on this challenge, it is first necessary to be able to grow cancer cells from a large number of cancers in the lab. For many years cancer drug discovery has relied on a small number of cell 'lines', derived from a cancer that was never sequenced. The cell lines undergo mutations and other changes during many years of culture on plastic and may be quite different from the cancer they came from. A new generation of cancer cell lines is needed. This is being achieved by growing cells in new ways in three-dimensional gels (see Fig A). Each cell line and the cancer it came from is genome sequenced to build up a collection of lines that represent all the common driver mutations in each cancer type.

Fig. A New cancer cell lines are being grown in specialized three-dimensional gels, forming organoids which are proving very useful for research.

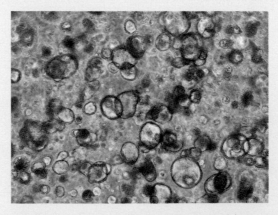

CGAP organoid derivation team, Wellcome Sanger Institiute, GRL

Once lines have been generated, genome editing using CRISPR–Cas9 (see chapter 7 for a clear description of this important technique) can be used to delete every gene in the genome to define which genes are critical for the survival of that cancer cell. This technology expresses a bacterial enzyme called Cas9, which is guided to a particular sequence in the genome by a short RNA fragment, so that a specific gene can be targeted. Once bound, the enzyme generates a double-strand break. After the repair of the break, the DNA sequence may be altered, disrupting the targeted gene. Sequencing is used to confirm that the targeted gene has been deleted. This can be applied to each gene in the genome, in each cancer cell line, to find which genes the cancer cells rely on to survive. Knowing which genes to target can guide the selection of anti-cancer drugs, which can also be tested on the cell line. This is a grand vision that will take years to deliver, but it points to a future in which each patient's cancer will be sequenced, and this will be used to select treatments based on data from the vast data sets of the vulnerabilities and effective treatments identified from cancer cell lines.

Chapter summary

- Cancer is a disease in which genomic changes in cells cause them to grow out of control.
- The genome changes in cancers include single nucleotide changes, amplifications or deletions of regions of chromosomes, and chromosome rearrangements that may join genes together.
- Some of the genome changes are 'drivers' that cause cancers to develop.
- Factors in the environment that alter DNA, such as sunlight in skin cancer, leave specific mutational signatures in the genome.
- We are protected from cancer development by genomic quality controls such as cell cycle checkpoints.
- Genomics is helping to discover new cancer treatments and guide which treatments are best for individual patients.

Further reading

1. The Mutographs project: https://www.mutographs.org/ mutographs-project/#mutograph.
2. What a mutational signature looks like: https://www.mutographs. org/2018/09/07/what-does-a-mutational-signature-look-like/.

biologically related to someone whose DNA is contained within a national database. Given that genomic information links us to our relatives, the decisions that we make about it (whether to donate it for research, whether to be tested, whether to withhold it) will all have an impact on those we are related to and the knowledge that they too can gain. It is this fact that makes genetic information quite different from other sorts of medical information. Thus, we all have a stake in how we as a society use genomic data (see Fig 3.1).

Fig. 3.1 The genetic information held in families and friends contains so much information about relationships, diseases, and more—should it be shared, or secret?

Anthony Short

People as patients

The UK has been at the forefront of using genomics in a health-care setting. In 2014, the Deciphering Developmental Disorders (DDD) study reported the first results from the sequencing of 12,000 children with developmental disorders. At the same time, work began on a project (the 100k Genome Project) to sequence 100,000 genomes of NHS patients. Outside of the UK, sequencing in health-care settings is also expanding, with investment in genomics seen on an international scale. In Europe, there is the 'Million European Genomes Project', where genomic and health data from 1 million

European patients will be linked. Iceland has a tiny population of under 350,000 and has a more complete genomic database of all its citizens than any other country. In the USA, the National Institute of Health Precision Medicine Initiative offers 1 million people some level of genomic sequencing. In Australia, the Genomics Health Alliance is currently creating the infrastructure to integrate genomic medicine into health care nationally. Possibly the most striking international example comes from Qatar, where the government has plans to offer genome sequencing to its entire population.

The pace of change is so fast that it is likely that even in the months between the writing of this book and its publication new genome sequencing projects will have emerged. What is clear is that genomic technology is now being used globally to provide diagnostic, prognostic, and treatment information for patients using health-care services in a way that has never been seen before. Many people, including eminent doctors and scientists, predict that the ability to look into the DNA of patients will transform health care.

The use of genetic testing in health care is not new. Currently, genetics is used in a wide range of health-care settings, including prenatal care (testing in pregnancy), oncology (cancer—see chapter 2), cardiology, ophthalmology, dermatology, and in the diagnosis of rare diseases (see chapter 1), to name just a few. Testing in these settings will continue. However, genetic testing (looking at single genes) is shifting to genomic testing (many genes), and the ability to test is becoming faster and cheaper. As a result, there is a dramatic increase in the amount of information generated. The genomic dream is dreamt on a large scale. As an example, as recently as five years ago a patient who developed breast cancer and also had family history of young-onset breast cancer would probably have been tested for two genes—BRCA1 and BRCA2. Mutations in these genes are known to lead to an increased risk for certain types of cancer, most notably breast and ovarian cancer. Now, a similar patient might well have a 'panel test' where over twenty genes associated with breast cancer are explored. Each of these genes will have different links to cancer, thus increasing the complexity of the results that emerge—but also helping doctors to choose the most effective treatment (see Fig 3.2).

Not only can we use genomic technology to understand the genetic basis of an existing condition, such as breast cancer, but it can also be used to uncover a new diagnosis for a previously undescribed rare condition. Again, the key change here is the volume of information that can be generated. Instead of testing just a few genes, a child with an unknown developmental disorder may now have 20,000+ genes sequenced and then filtered using bioinformatics to explore all of the genes linked to intellectual disability, autism, and developmental conditions. This may reveal new diagnostic information, telling us what is wrong, or prognostic information which helps to explain what may happen in the future as the condition progresses.

Whilst it is unrealistic to suggest that all 20,000 genes will be analysed and reported as a whole, the resource is at least available to be interrogated as and when required. As genomic medicine is increasingly available across

Fig. 3.2 Mammograms screen for breast cancer which has already started to develop. Genome testing makes it possible to predict who is at high risk of the disease—perhaps enabling them to be screened more often.

Tyler Olson/Shutterstock.com

health-care systems, we are moving closer to genetics becoming used routinely in a range of different health-care settings, sometimes referred to as genetics being 'mainstreamed'. This mainstreaming means that patients will have more exposure than ever before to large volumes of complex genetic information.

The increasing prevalence of genetics in health care raises some important ethical questions, such as how genomic technology should be used. This is important, as there is evidence that using genomic medicine does not always simplify or improve care. Take the two examples above. In the case of rare diseases, research has shown that implementing genomics can increase diagnostic uncertainty; it is not always clear that the genetic alteration identified is the cause of a person's rare disease. As the scale of genetic testing dramatically increases, this—inevitably—also means an increased uncertainty in many of the results returned. Many genetic results can only be interpreted as ambiguous or uncertain, because we still know and understand so little in this field. Such 'variants of uncertain significance' are results where the meaning is unclear and they are more likely to be discovered when multiple genes are tested for at the same time.

In breast cancer there is also evidence that genomic medicine has, in some cases, complicated and, importantly, not improved treatment pathways. So, while the potential benefits of genomics are great, it is important not to get carried away with rhetoric and view genomics as a magic cure for all ills; genomics will not be a universal panacea. Researchers and health-care professionals must be open, at least in principle, to the idea that more genomics might not always be better—at least until we understand more about what is going on!

Now we have the ability to look at multiple genes within one test, genomic technologies give an opportunity to explore genes unrelated to the health condition that a patient was originally referred for—a piece of serendipity or good luck! This means that, for example, a patient might have a diagnosis and family history of cancer genes. As such, they might well be tested for a number of genes linked to an increased risk of cancer. However, at the same time it would also be possible explore their genes linked to other conditions, such as heart disease or type 2 diabetes. Such results are referred to as 'additional looked for findings', 'secondary or incidental findings', or sometimes as an 'opportunistic screen'.

Whatever they're called, the basic principle is the same: when genetic testing is carried out on an individual, it is likely that more information will be available than that relating to the original diagnosis. We can see this trend in the NHS's 100,000 Genomes Project (which finished recruitment in 2018). Here, children who have a rare disease where there is believed to be a genetic cause are offered genome-wide testing. In this clinical/research study, the parents of each child having genome sequencing also have the opportunity to be tested for 'additional looked for findings' related to their own risk of future disease, unrelated to their child's condition.

What does all this testing mean?

As a result of this data explosion, patients will be faced with complex choices about what they want to know about the information hidden in their genome. As patients are exposed to increasingly complex and uncertain genetic information, debates have emerged as to the best way to manage this. Research demonstrates that many people are enthusiastic about receiving genetic information. However, health-care professionals (especially those who work in genetics) are often more cautious. Many who work in genetics have voiced concerns that too much genetic information could do more harm than good. Some have argued that the best way to view your genome is to see it as a resource that should be accessed in different ways over the lifetime of an individual. In other words, there should not be one moment when you find out everything there is to know about your genome. However, patient groups, social scientists, and health-care professionals are still debating the best way to manage the increasing volume of complexity and uncertainty generated by genomic tests (see Bigger Picture 3.1).

A key aspect of genomic information that makes it different to other sorts of medical information is that it is shared between biological relatives. So, even if a person is not using health-care services themselves, they may be related to someone who is; thus, the reach of a genetic result moves outside of the clinical encounter and into the wider family. Whilst most of us will not currently be a patient in a health-care setting, we may still be related to someone who is having testing and the questions they have answered may be very relevant to us too. Therefore, the impact of genomic information naturally extends beyond a health-care encounter—through conversation it travels from the patient, out to their extended family, and to people who are not yet patients. Such people could be considered a type of 'patient in waiting'. We don't yet know where and how people who have never

Bigger picture 3.1
To test or not to test: Huntington's disease

Because we *can* do something, it doesn't always follow that we *should*.

Take the case of Huntington's disease, which affects 6–10,000 people in the UK alone. This disease is caused by a dominant mutation or pathogenic variant, in a gene on chromosome 4, often referred to as the HTT or huntingtin gene (see Fig A). In healthy individuals, this gene codes for the production of huntingtin, a protein which seems to be needed by the neurones in the brain, to enable them to function properly, and by the mitochondria. The most common form of Huntington's disease is a progressive brain disorder where the symptoms appear when the individual is in their thirties or forties. It causes a wide range of symptoms, from irritability, poor coordination and problems learning new things, to involuntary twitching and loss of control of movements, making walking, talking, and even swallowing difficult. Eventually, over a period of between fifteen and twenty years, it is fatal.

Huntington's disease is caused by a repeat expansion mutation in an area of the HTT gene with a CAG trinucleotide repeat, a sequence which results in extra glutamines being added to the final protein (Fig A). In healthy people, the CAG segment is repeated between ten and thirty-five times within the gene. When the mutation occurs, people may have between thirty-six and 120

Fig. A The mutation in the HTT gene that results in Huntington's disease.

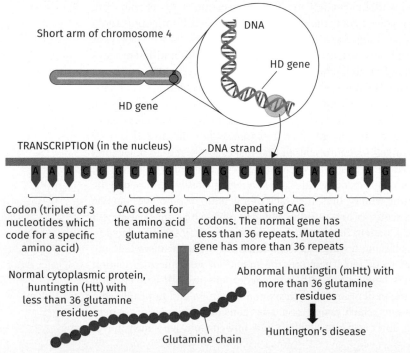

© Blamb/Shutterstock.com

repeats. People with more than forty repeats will always be affected by the disease. They produce an abnormally long version of the huntingtin protein. This is then cut up into shorter fragments in the body, and these fragments are toxic to the neurones where they accumulate. The neurones stop functioning properly and eventually die, and this death of brain neurones leads to the signs and symptoms of the disease.

Because Huntington's disease does not usually develop until an individual is in their late thirties at the earliest, many people have already had children before they realize they are affected by the pathogenic variant. Now, however, there are genetic tests which can be used if a member of a family is diagnosed with Huntington's disease. So, for example, the siblings or children of an affected individual can discover whether or not they have the dominant mutation of the *HTT* gene that means they will inevitably develop the disease.

There is currently no effective cure or treatment for Huntington's, so although it is perfectly possible to test individuals for the *HTT* mutation, not everyone wants to have the test. There are a number of personal and ethical issues involved in these decisions—for example:

- Once you know you have the *HTT* mutation, you know that you will develop a fatal disease. This affects many life choices—marriage, having children, work patterns, finances. Some people want as much information as possible. Others would rather not know how the genetic dice have fallen.
- Having a test for Huntington's does not affect one individual alone. If one person in a family is shown to have the mutation, their siblings may also have the faulty gene, and so may their children and grandchildren. Some people may want to know that they are at risk, to plan their lives accordingly. Others may find the information a very heavy burden and may prefer not to be aware of potential risks.

❓ Pause for thought

1. Think of as many reasons as you can for and against having a genetic test for an incurable disease such as Huntington's.
2. Scientists are currently trialling a new type of gene-silencing drug, RG6042. It is an antisense oligonucleotide which is designed to target and destroy all forms of mutant *HTT*. This drug will silence the *HTT* gene, lowering the levels of mutant huntingtin. One of many exciting things about this approach is that the drug doesn't simply treat symptoms—if it works as scientists hope, it will actually affect the underlying disease process, preventing the toxic protein from forming. The trial will show whether the new drug halts progression of the disease and has an impact on the symptoms experienced by those affected.

   However, in its current form the drug also knocks out normal huntingtin. Although animal trials and initial human trials suggest that the drug is safe, there is still a long way to go.

   Consider:

   a) the ethics of trialling a drug that affects both healthy and mutant huntingtin levels;
   b) how the availability of an effective treatment for Huntington's disease might affect attitudes to genetic testing for the condition, and why.

encountered genetic testing before will then seek out information and make sense of it, but research suggests that people will look to the media, popular culture, and the internet for insight.

As many different people begin to be faced with genetic information in health-care settings, an important issue to tackle is how to accommodate different opinions and attitudes. One issue that arises when studying genetics is what non-experts make of genetic information. The evidence we have, especially from genetic counselling, suggests that the ways people understand and make sense of genetic information are complex, varied, and often idiosyncratic. An important ethical issue is who 'owns' a piece of genetic information. Is it the individual's information, or does it belong to the family? As an example, if an individual is found to have genetic alteration (say in *BRCA1*), then other family members could benefit from this knowledge. People with alterations in *BRCA1* are offered various different interventions to manage their increased risk of cancer, such as increased screening. But what if someone doesn't want to share that information? Do health-care professionals have a duty to warn other family members about this (see Fig 3.3) or should individual confidentiality be respected? In genomics, there are important questions to be answered about the rights and responsibilities of patients who have had testing and the health-care professionals who care for them.

The complexity of these issues means that the provision of genetic counselling is of increasing significance. Patients have voiced the opinion that the provision of genetic counselling is important when having genome sequencing. Researchers and clinicians have also highlighted the importance of drawing on the expertise of genetic counsellors in order to ensure that new technologies are integrated appropriately into health care. Patients

Fig. 3.3 Genetic testing carried out on one of these siblings could have serious implications for any of the others—or their children or future grandchildren.

Ann Short

have voiced the opinion that the provision of genetic counselling is important when having genome sequencing. Researchers and clinicians have also
highlighted the importance of drawing on the expertise of genetic counsellors in order to ensure that new technologies are integrated appropriately
into health care. The problem is that, however desirable counselling may
be, it is expensive, and stretched resources may not be used to provide it.

People as consumers

Having discussed some of the implications of genetics for patients, we now
explore some of the ways that genomics will affect people as 'consumers'.
One way is via 'direct-to-consumer' (DTC) genetic testing. This is a growing
industry, with private companies marketing and selling a wide range of
tests through the internet (see Fig 3.4). Consumers are able to send off a
DNA sample (normally a saliva sample) and a few weeks later receive their
test results.

A wide range of DTC genetic tests exist, both health-related and non-
health-related. Health-related tests include identifying predispositions to
common and complex disorders, such as cancer and cardiovascular (heart)
disease. They also include tests to determine whether people are 'carriers'
for rare recessive genetic disorders. While people will not be affected by
these conditions, this information could affect their children, and so might
guide reproductive decisions. Still further, these tests can provide information about how well an individual can metabolize drugs—'pharmacog-
enomics'—and how diet interacts with your genome—'nutrigenomics'. All
this information could potentially guide treatment and lifestyle choices.

Non-health-related tests include testing for paternity, ancestry, and
athletic ability, as well as traits such as earwax characteristics and caf-
feine metabolism! DTC genetic testing is rising in prominence. In 2008,
23andMe's retail DNA testing kit was named invention of the year by *Time*
magazine. In 2016, genomics was named by Forbes as one of the three
'Big Technologies to Watch' over the subsequent decade, together with

Fig. 3.4 DNA testing kits you can do at home may seem like harmless fun
or the chance to gain insights which may help to keep you healthy—but
might you discover more than you bargained for?

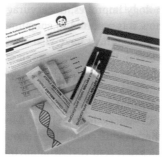

nevodka/Shutterstock.com

6. Direct-to-consumer testing: https://www.yourgenome.org/stories/direct-to-consumer-testing.

7. What genetic counsellors do: https://www.yourgenome.org/stories/genetic-counselling.

8. The UK National DNA Database: https://www.yourgenome.org/facts/what-is-the-uk-national-dna-database.

9. The ethics of having a DNA database: https://www.yourgenome.org/debates/is-it-ethical-to-have-a-national-dna-database.

10. Genetic modification of farm animals: https://www.yourgenome.org/debates/is-it-ethical-to-genetically-modify-farm-animals-for-agriculture.

11. Genomics and crime: https://www.yourgenome.org/stories/the-eureka-moment-that-revolutionised-crime-solving.

12. The Christmas present that could tear your family apart: https://www.bbc.co.uk/news/stories-46600325.

Discussion questions

3.1 How important is it to ensure that people have access to new genetic technology?

3.2 Discuss the pros and cons of people having their genomes sequenced.

3.3 Suggest, with reasons, the most appropriate time for genetic testing to be carried out—for example, at birth, on reaching adulthood, before starting a family . . .

3.4 Direct-to-consumer genetic testing raises many questions. Suggest some of the problems which might arise as a result of it. Should genetic counselling be compulsory?

3.5 In December 2018, there were unconfirmed reports that the first genome-edited babies were born in China. This was met with international condemnation. The scientist who claims to have carried out the procedure is now in prison. Why do you think the response was so negative—and how do you think this will influence policies over the future use of genome editing in the quest for a healthier human population?

4 PATHOGEN GENOMICS

The modern science of genomics really began with pathogen genomics. As scientists developed sequencing tools at the end of the 1970s, virus genomes were the first to be tackled, because of their small size. Viruses, and then bacteria, formed the test bed for developments that went on to enable the sequencing of human genomes, but they are also of interest, and of value, in themselves.

As you have discovered, a genome is a complete parts list for an organism, so a pathogen genome tells us a lot about how pathogens grow and interact with their environment. This environment may well include us! The genome also tells us how the pathogens avoid both our immune systems and the weapons we have developed to overcome them, such as vaccines and antibiotics.

Genomics shows us that bacterial genomes are very dynamic. Unlike eukaryotes, which only inherit DNA from their ancestors, pathogens can acquire DNA from other organisms living around them, keeping what is useful and adding it to their own chromosomes. Genomics demonstrates how this plays out, and also reveals the mechanisms pathogens use to specifically accelerate the variation of their genomes to allow them to adapt rapidly to new environments. Figure 4.1 shows you *Neisseria meningitidis*, a disease-causing bacterium, and its genome.

Beyond this fundamental information, a genome is also a written record of the evolutionary history of the pathogen. When read carefully, and compared to genomes from close and distant relatives, we can reconstruct this history. We can see when, and where, particular pathogens evolved and how they spread across the world.

Fig. 4.2 The complete genome of a free-living organism—the bacterium *Haemophilus influenzae*. The coloured ticks around the outside circle represent the genes, colour-coded by function. The other features represent features of the genome, or the sequencing approach.

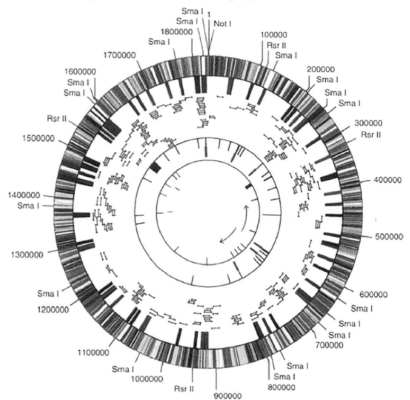

RD Fleischmann, MD Adams, O White, RA Clayton, EF Kirkness, AR Kerlavage, CJ Bult, JF Tomb, BA Dougherty, JM Merrick, et al. Whole-genome random sequencing and assembly of Haemophilus influenzae Rd. *Science*. The American Association for the Advancement of Science. Jul 28, 1995 Copyright © 1995

Over the next few years, genomes from many of the major pathogens were sequenced, including those causing cholera, syphilis, meningitis, tuberculosis, dysentery, food poisoning, plague, and typhoid fever. At that time, each of these was a major undertaking, often costing hundreds of thousands of pounds and taking teams of people many months to complete, so strains for sequencing were chosen carefully, and some effort was taken to avoid duplication. Later, some groups started to perform comparative genomics, sequencing two or three strains from the same species, such as *Helicobacter pylori* (which causes stomach ulcers) or *E. coli*, but again this was limited due to cost.

All of this changed dramatically with the introduction of highly parallel approaches to sequencing, often called next-generation sequencing (NGS), which is highly automated and very fast (look at chapter 7 for more about this process). The development of NGS dropped the cost and time of sequencing so much that it became possible to sequence the genomes

Fig. 4.3 This phylogenetic tree (a) shows the ancestry of strains of *Staphylococcus aureus* (b). Each branch represents a strain, and the links between the branches (nodes) represent the ancestor of those two strains, back to the single common ancestor (CA) of all the strains. The labels in boxes represent the acquisition of different antibiotic resistances. The colours of the branches show the countries the strains have been isolated from.

(a)

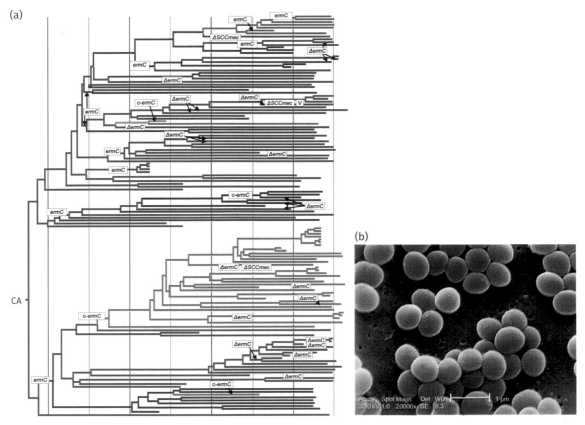

(b)

(a) Adapted from Holden, MTG et al. 'A genomic portrait of the emergence, evolution, and global spread of a methicillin-resistant Staphylococcus aureus pandemic.' *Genome Research* 2013. 23: 653–64. © 2013. Published by Cold Spring Harbor Laboratory Press. Doi: 10.1101/gr.147710.112. CC BY NC 4.0 (b) CDC/Matthew J. Arduino, DRPH

of large numbers of very closely related genomes. The effects of this step-change in capability were huge. Comparing closely related genomes allows you to build up a phylogeny, or genealogy, to identify how pathogens are related, and to trace them back all the way to a single last common ancestor (LCA). You can see an example in Fig 4.3. This also reveals the changes that each has made as they have evolved into their current state. Combining this with information on the time and place that each strain was isolated allows you to infer the location and time at which each of the ancestors existed.

These pathogen phylogenies can be tremendously powerful. They can help us identify the point of origin and patterns of global spread of recently evolved pathogens, and allow us to identify when and where drug-resistant lineages arise. Pathogen phylogenies enable us to trace the evolution of

Fig. 4.4 The exceptions that prove the rule! There is nothing in the genome of a cow to indicate it can survive—and produce calves and milk—based on a diet of grass. For that you need to explore the genomes of the bacteria that live in the gut of the cow!

© Anthony Short

cytoplasm, cannot be created from scratch, and must be inherited directly from the parent cell. As another example, you could understand everything about the genome of a cow, and still not be able to infer that it eats grass (see Fig 4.4), because it does not encode the enzymes that allow it to break down cellulose, the primary energy source in grass. In fact, these enzymes are produced by bacteria that live in the cow's rumen. Similarly, host-obligate pathogens (those that cannot survive outside the host), such as *Mycoplasma* or *Chlamydia*, usually cannot make their own nucleotides and amino acids, as they scavenge these from their host. But apart from a few cautionary counterexamples like these, the majority of an organism's specific functions are encoded in their genomes, and this is especially true of single-celled bacterial pathogens (and viruses).

Of course, the genome itself is just a string of several million As, Cs, Gs, and Ts, and in its raw form makes very little sense. Having generated this long string of apparently random letters, the next step is to work out which bits of the sequence code for genes.

Bacterial genomes tend to have about one gene every thousand base-pairs (kb), each of which is just under a kb long, on average. This means a typical bacterial 4 million base-pair (Mb) genome will have about 4,000 genes. There are a number of different computer programs designed to find these genes, and most work by looking for signals in DNA. There are some obvious signals, like ATG to start translating a gene, or TAA to stop, but there are a lot of random ATGs and TAAs, and not every space between them encodes a gene. The program must therefore look for more subtle signals, often using the base composition, or signals for the start and stop of transcription.

Having predicted the location of the genes, we must then try to predict their function. This is done by comparing them to databases of genes whose function has already been worked out, and assuming that similar proteins will have similar functions in different organisms. Ideally, this functional determination will have been done experimentally, with a clear link between a gene and the protein function shown. But as more and more genomes have been sequenced, the databases have filled up with genes whose function has been inferred by comparison with other genes, leading to long chains of inference based on a few experimentally derived functions. Add to this the problem that many genes have never had their function identified at all, and you arrive at the dirty secret of genomics: in many cases, we really don't have a clue what the genes are doing, even though we know they are there!

In the early days of pathogen genomics, these parts lists (imperfect as they were) were scrutinized in great detail to discover everything possible about the biology of the organism, particularly how it interacted with the host and caused disease. Often the first step in this process is metabolic reconstruction, where all the genes that code for enzymes are added into a model of the metabolism of the cell. This often allows us to identify what nutrients the organism needs for growth, and in some cases it has allowed us to design growth media for previously unculturable bacteria. Perhaps more importantly for pathogens, we can also look for genes that are involved in interactions with the host, to help us understand more about how the organism causes disease. These can be obvious, such as genes that make toxins, or sets of genes that make surface fibres for sticking to host cells. Bacterial pathogen genomes also encode complex machines such as the type-III secretion systems, which are used to inject proteins directly into the host cytoplasm to take control of the host cell. These can be readily identified, along with the catalogue of secreted proteins.

Most often, these discoveries will not answer questions themselves, but will generate hypotheses and ideas for further experimentation. One of the strengths of genomics is its ability to generate broad, novel hypotheses for testing that are not based on the limited previous experimental knowledge of the pathogen.

Something that becomes obvious from the analysis of pathogen genomes is that these pathogenicity genes are often clustered into distinct sections of the genome that are absent from non-pathogenic relatives. These sections also display hallmarks of horizontal acquisition; that is, at some point they have been acquired through the acquisition of DNA that was not from the direct parent of the cell. Bacteria have many ways of acquiring DNA, and can incorporate this DNA directly into their genomes where it will be passed on to their descendants. More of this later! These clusters of genes are called pathogenicity islands (see Fig 4.5), and they can be recognized by several features, such as having unusual nucleotide composition, and being inserted at particular points in the chromosome. We can turn this discovery process round and look for new pathogenicity islands using these features. This then tells us that the unknown genes in these novel regions

Fig. 4.13 Colonies of *Streptomyces coelicolor* bacteria overproducing the blue antibiotic actinorhodin, seen as droplets on the surface of the colony

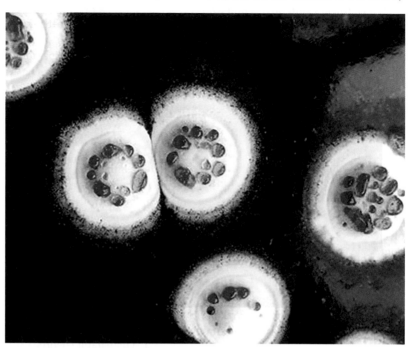

© Paul Hoskisson and Matt Hutchings, University of East Anglia/Norwich Research Park (NRP) Image Library/CC BY 4.0

have not yet been successful in identifying significant novel antibiotics using this approach. They have largely fallen back on screening against whole pathogens, rather than isolated targets, although the approach is still being used in academic labs.

Despite this apparent failure, genomics is still valuable in antibiotic development. One area of particular value is the identification of antibiotic targets and modes of action. Identifying a novel candidate antibiotic using a whole-cell screen does not provide information on how that antibiotic works, or the target in the cell. This data is necessary for further development and clinical use of the antibiotic. Selecting resistant bacteria in the lab and sequencing their genomes will rapidly identify the genomic changes that have led to resistance, often identifying the target of the antibiotic within the cell. This process of genomic sequencing of resistant strains can also be used on pathogens that have developed resistance in the wild, which can be due to point mutations, or acquisition of mobile resistance genes. Understanding how bacteria become resistant, and how they exchange that resistance amongst themselves, can lead to better antibiotics, or to the development of new chemicals that target the resistance mechanism itself.

Finally, being able to recognize, from the genome sequence, all the mechanisms that cause resistance means that we can predict the resistance of a pathogen directly from its genome sequence. This is currently being used in the clinic. All patients with HIV in the UK have the genes from the pathogen sequenced whenever their medication is updated or changed. Comparing these sequences to databases of known resistance variants allows doctors to determine whether the treatment is likely to work or not, and the most effective medication for each patient can therefore be chosen.

The same process is now being trialled for *Mycobacterium tuberculosis*. Tuberculosis is a major global threat, but the bacteria that cause the disease are becoming increasingly resistant to commonly used antibiotics. Because *M. tuberculosis* takes so long to grow in the lab, testing bacteria from a particular patient for drug resistance can take many months, during which time the patient may be being treated with antibiotics that are ineffective against the strain they are carrying. Sequencing the bacterial genome on diagnosis allows us to predict resistance much more rapidly, and so choose the correct and most effective treatment right at the beginning of the treatment process. It is likely that, as sequencing technologies get cheaper and faster, this approach will be extended to many other pathogens.

Outbreak detection

The final area where genomics is of increasing importance in the clinic is an extension of the phylogenetic approaches used for studying the historical emergence and transmission of pathogens. This can be at the local level, looking at pathogen transmission in a single hospital, at a country-wide level, particularly when looking for foodborne pathogens, or at an intercountry level, trying to understand urgent outbreaks. These techniques apply to both human and veterinary medicine. We will look at examples of each of these.

In an earlier section, 'Dating pathogens', we examined one particular MRSA strain. ST22 emerged in the UK during the late 1990s, sweeping through hospitals and causing large outbreaks and significant disease. This outbreak has subsided, but transmission of MRSA still occurs in hospitals, and they need to be able to identify when transmission between patients has occurred in order to intervene.

The problem is that all ST22 look identical when using normal typing techniques, so you cannot say whether two patients in the same ward carrying ST22 have given it to each other or brought it in to the hospital independently. The previous analysis of the clock rate in *S. aureus* showed that it mutates about once every six weeks or so, generating enough variation to easily differentiate different ST22 from each other, and even to track the path of transmission from patient to patient. This approach has already been used in hospitals to intervene in MRSA outbreaks (see Case study 4.1—MRSA on the Special Care Baby Unit), and is being expanded to other important hospital pathogens, such as *E. coli*, *C. difficile*, and *Klebsiella*.

Case study 4.1
MRSA on the Special Care Baby Unit

Staphylococcus aureus is a common commensal bacterium, carried on the skin or in the nose of about one-third of the population at any time. MRSA is an antibiotic-resistant type that can be carried by about 1 per cent of the general population, but is more common in hospitals. Although it is a commensal in healthy people, it can cause life-threatening infections in hospitalized people who are already unwell—and MRSA is hard to treat, due to its resistance. This is why it is important to identify patients carrying MRSA, and even more important to find out whether they are transmitting the bacteria to others. As described earlier in this section, current typing techniques do not provide sufficient resolution to differentiate between patients independently bringing the same strain into the hospital, and infections due to transmission between patients.

In 2011, a group of researchers (including the author) was trying to show that whole-genome sequencing (WGS) had sufficient resolution to be used in hospital to link patients to MRSA outbreaks on the wards, and also to exclude them from those outbreaks when they were not involved. To test this out on a real-life example, the scientists looked at a cluster of MRSA-positive babies who had been detected on the Special Care Baby Unit (SCBU) a few months earlier. Although this cluster had been investigated by the infection control team at the hospital, a number of features, such as gaps in the apparent transmission chain, meant that they could not be certain it was an outbreak. The health-care professionals instigated enhanced infection control procedures, and there were no more cases.

The researchers dug the MRSA isolates out of the freezer and sequenced them, and were able to show that there was indeed an outbreak. They then went to the freezers and looked for other isolates that could be part of the outbreak. They identified more from babies on the ward that were missed at the time, from mothers on the associated maternity ward, and from the partners of those mothers in the community. While they were doing this retrospective analysis, the hospital contacted them, and said that there was another baby on the ward with MRSA, two months after the outbreak had apparently stopped. Was this baby part of the outbreak, and was it still going on? The researchers used a new sequencing machine that gave data in a much shorter time, and were able to confirm within a few days that the baby was linked to the earlier outbreak. The long gap with no MRSA-positive baby on the ward suggested that the MRSA strain was being carried, not by patients or visitors but by a health-care worker on the ward.

The hospital asked all the workers associated with the ward to volunteer to be screened for MRSA and all 150 accepted. From them, one screened positive for MRSA. This was not conclusive, as you will remember that 1 per cent of people carry MRSA anyway (and this frequency is higher in health-care workers), so this could have been an unrelated strain. The researchers again sequenced the isolate using the rapid-turnaround machine, and were able

Fig. A New-born babies are vulnerable to infections. Whole-genome sequencing made it possible to protect them in this case—with ever smaller, faster, and cheaper sequencing machines this type of intervention will play an ever-increasing role in medicine in the future.

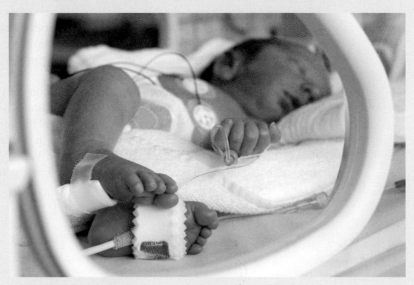

iStock.com/metinkiyak

to link it directly to the outbreak, showing that the health-care worker had probably reintroduced the strain onto the ward after the gap. The worker was taken off the ward, and treated to stop their MRSA carriage. After this, there were no further infections on the ward with that strain. This was the first time ever that whole-genome sequencing had been used to intervene in an ongoing outbreak in a hospital, preventing the distress of further infections and even the potential loss of life (see Fig A).

❓ Pause for thought

Should we screen hospital staff for MRSA? Hospital staff have a higher carriage rate for MRSA than the general public, but it is not clear how often they are involved in transmission to patients, compared to transmission directly between patients. What would be the potential for damage to the careers and/or the mental well-being of staff identified as carriers? How would the cost of such an approach compare to its benefit?

Outbreaks of foodborne pathogens, such as strains of *Salmonella* or *E. coli*, can be particularly difficult to track. In a modern economy, the food on your plate can come from many different sources, both national and international, and food producers will distribute their products across very large distances. In this situation, linking up cases of *Salmonella* infection across a whole country and identifying the source of the outbreak can be very difficult, especially as it will often involve different public health investigators. Again, genome sequencing provides sufficient information to unambiguously link cases to each other, and to the source. Perhaps even more importantly, the genomic data is digital, and therefore easily exchanged and compared across large distances using centralized databases. Genome sequencing is already being used by Public Health England in the UK and the Food and Drug Administration in the USA to perform routine outbreak detection.

Genome sequencing is often thought of as difficult and complex, and therefore applicable only to first-world problems. Although this may have been true a few years ago, newer portable and rapid-sequencing devices have been developed that allow sequencing directly in the field. These have already been used to help in an Ebola outbreak in West Africa (see Case study 4.2), and this kind of technology is likely to bring genomics directly to the patient bedside in the future.

Case study 4.2
Ebola sequencing in the field

In December 2013, an outbreak of Ebola started in West Africa. It went on to become the most widespread outbreak in history, compounded by poor health infrastructure in the affected countries, and the final death toll exceeded 11,000. As the disease became established, and as part of much wider (if somewhat belated) responses to the epidemic, two groups sought to discover whether WGS could play a part in trying to bring the outbreak under control.

One group, based in Cambridge, used a standard benchtop sequencer (an Ion Torrent) that had to be set up and validated in a lab in the UK, before being shipped to Sierra Leone on a humanitarian aid flight. The equipment included reagents for sequencing, unassembled benches, polymerase chain reaction (PCR) cabinets, centrifuges, general molecular biology reagents, N_2 canisters (required for the sequencing machine), a sequencing library preparation robot, and the sequencing machine itself. The system was installed in an air-conditioned tent adjacent to a Public Health England-run diagnostics facility. Once there, the scientists had to contend with high temperature, dust, high humidity, sporadic power supplies, and poor internet access, the last important as the sequences had to be sent back to the UK for analysis. Despite this, the group managed to generate genome sequences for 554 Ebola isolates, with about a four-day turnaround time, representing nearly 24 per cent of the cases in Sierra Leone in 2014.

The second group took a more minimal approach. They used an experimental sequencing device (a MinION), that is about the size of a chocolate bar (see Fig A). It plugs directly into the USB port of a standard laptop, drawing

Fig. A The MinION system for sequencing DNA in action in Sierra Leone.

© Dr Sophie Durrafour.

its power from, and delivering its data to, the computer. Although they still required PCR machines and other reagents, they managed to pack the entire kit into standard airline passenger luggage. This group set up their laboratory in Guinea, sequencing 142 samples with a turnaround time of about two days. This group had to contend with similar problems to the first, although power continuity was less of an issue as the laptops could run off their own batteries. Again, the data had to be sent back to the UK for analysis over a poor internet connection. In spite of internet problems, this is still considerably easier and quicker than sending the samples themselves for sequencing.

What did genome sequencing achieve in the Ebola outbreak? The sequencing labs arrived late in the outbreak, when the epidemic was already well established, so it was unable to have an effect at this stage. Using phylogenetic approaches, a combination of the data sets was able to show the routes by which the epidemic had transmitted within and between countries. In addition, sequencing did turn out to be very valuable in 'mopping up' towards the end of the outbreak, tracing sources for residual cases that popped up in odd places and helping efforts to control these. It was also able to show that some of these cases were due to previously unknown mechanisms of transmission, such as sexual activity and breast milk, and that recovered victims could carry the virus for much longer than previously thought.

These interventions were able to show the value of WGS even in resource-poor settings, pointing the way for future use of more easily deployed technologies as they are developed.

Pause for thought

This was an intervention by researchers and public health workers from the developed world. What would be required to allow this to happen routinely in low-income countries? Would this be useful (or cost-effective) in countries that lack basic health infrastructure?

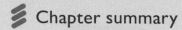 Chapter summary

- The tools and techniques of genomics were developed and honed on pathogens.
- Genomics generates a parts list for pathogens that can be used to understand pathogenicity.
- Bacterial genomes vary by more than just mutations. Acquisition and loss of genes over time leads to a shared genome (the pan-genome), which is much larger than the number of genes in any individual member of the species.
- Genomes represent a written record of the history and evolution of the pathogen, which can be used to identify emergence, spread, and adaption of the organism.
- Pathogen genomes encode mechanisms that generate variation in a programmed way in specific sets of genes.
- Bacteria often act as part of larger communities (microbiota) and sequencing the total DNA of this community (metagenomics) helps us study this.
- Pathogen genomics has clinical applications in vaccinology, antibiotic development, and outbreak detection.

Further reading

1. Targeting disease: https://www.yourgenome.org/topic/targeting-disease.
2. The fight against cholera: https://www.yourgenome.org/stories/science-in-the-time-of-cholera.
3. Superbugs: https://www.yourgenome.org/stories/tracking-superbugs.
4. Using DNA to track disease outbreaks: http://theconversation.com/genetic-detectives-how-scientists-use-dna-to-track-disease-outbreaks-57462.
5. A review on antibiotic resistance: https://www.ncbi.nlm.nih.gov/pmc/articles/PMC5573035/pdf/cureus-0009-00000001403.pdf.
6. Tackling drug-resistant infections globally: https://amr-review.org/sites/default/files/160525_Final%20paper_with%20cover.pdf
7. An on-line course on bacterial genomics: https://www.futurelearn.com/courses/introduction-to-bacterial-genomics.
8. Video link on working with human gut microbiota: https://www.yourgenome.org/video/life-in-the-lab-working-with-human-gut-microbiota.

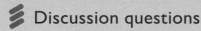 Discussion questions

4.1 Bacteria use many mechanisms to enhance their genomic variation. To what extent does this accelerate their evolution compared to eukaryotes? How does this affect host–pathogen interactions and what are the implications for the development of vaccines and antibiotics against these pathogens?

4.2 Can high-tech solutions to bacterial diagnosis, resistance detection, and transmission analysis actually be useful in low- and middle-income countries where spend on health care is very low? Investigate and explain the concept of leapfrogging technologies.

5 PARASITE GENOMICS

The idea of the human genome is widely recognized, if often poorly under-stood. Chapter 1 explains how we are using our knowledge of the human genome to understand both the genetic diseases which affect relatively large numbers of individuals around the world and some of the very rarest dis-eases too. As you discovered in chapter 2, scientists are increasingly look-ing into and understanding the genome changes that take place in diseases such as cancer. We are using these insights to help us diagnose and treat the many different forms of the disease more effectively. And in chapter 4 you will have seen how we are learning more all the time about the genomes of pathogens that cause so many infectious diseases—the bacteria and viruses that can make our lives a misery and that cause millions of deaths every year.

In this chapter, we look at how genomics is shedding light on a widely vary-ing group of organisms: parasites. We know that they can cause devastation and disease, but we are also beginning to realize that some parasites may play crucial roles in our health and well-being. Parasitic infections may have ben-eficial effects on our immune systems, and the possibility that parasites have the potential to deliver previously unrecognized benefits as well as causing devastating harm makes it even more important to understand their biol-ogy. In this chapter, you will be finding out about research into the genomes both of parasites and their hosts. Ranging from fleas and ticks to protists and worms, parasitology is an immense and fascinating field of biology— and the genomics of these amazing organisms have some fascinating stories to reveal (see Fig 5.1).

Fig. 5.1 A tick is an external parasite—this one is absolutely full of cat blood. And when they feed, they can transmit disease . . .

© Anthony Short

What is a parasite?

Most people living in countries like the UK have thankfully never suffered from a serious parasitic disease—they are certainly not a major concern in our lives. Of course, lots of school-age children in economically developed countries occasionally experience the unpleasantness of having nits or threadworms, but surely parasites aren't enough of a problem for us to want to spend time and money understanding parasite genomes and what (biologically) makes them tick (see Fig 5.1—pun intended!). But wait a minute—don't our pets and livestock in the UK suffer from parasites, such as worms? And aren't there some diseases caused by parasites that affect people in tropical countries? So maybe there are good reasons after all to understand a bit more about them . . .

Parasites are organisms that survive by living inside, or on, another organism (also known as a host) at the expense of that organism's health. There are parasites within all different taxonomic groups: they can be animals, plants, fungi, viruses, protozoa, or bacteria. The parasite depends on its host for nutrients, which it needs to survive. Parasites can also be manipulative, sometimes even changing the behaviour of their host to maximize the chances of the parasite reproducing and continuing its life cycle (see the story of *Toxoplasma gondii* in another publication in this series, *Human Infectious Disease and Public Health*). Parasites affect virtually every species on the planet—from fish, reptiles, birds, and mammals to crustaceans

Fig. 5.3 Genomics can help us understand parasitic diseases such as Leishmaniasis—and may help us find a cure

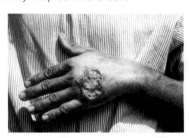

CDC/Dr D. S. Martin

The nucleotide content of parasite genomes also varies considerably. As discussed above, the *Plasmodium falciparum* genome has a very high AT content, with almost 80 per cent of the nucleotide bases being adenine or thymine, whilst other genomes have quite neutral GC content, for example that of the sheep parasite *Haemonchus contortus* (56 per cent GC).

Parasites have some interesting, quirky genome biology. Leishmaniasis is an example of disease caused by a protozoan parasite (a singled-celled parasite); it affects millions of people around the world, with symptoms including skin ulcers like those seen in Fig 5.3. There are at least twenty *Leishmania* species that cause disease. The protozoa are spread by the bite of infected female sand flies, which feed on blood to produce eggs—and there are over ninety species of sand flies that can spread the disease!

Between 20,000 and 30,000 people die of Leishmaniasis every year. So, anything that can help us tackle this disease is important.

Scientists have discovered that *Leishmania sp.* genomes have variable numbers of chromosomes; within an individual parasite some chromosomes are present more than once and some are missing entirely. This is known as aneuploidy. In most other organisms, aneuploidy causes severe abnormalities and is not well tolerated. An example of aneuploidy in humans is Down Syndrome, where a single extra chromosome 21 can cause a wide range of developmental and health problems. Interestingly, aneuploidy appears to confer an advantage in *Leishmania sp.*, perhaps by significantly altering expression levels of genes that enable the parasite to exploit specific host niches. Learning about the characteristics of parasite genomes can reveal much about how parasites have evolved their specialized way of living inside another organism.

Parasite genomes may contain interesting features from a genome biology perspective, but how exactly can we use that information to tackle parasitic diseases?

Tracking parasites

Genomics has been shown to effectively monitor the spread of bacterial and viral parasitic diseases, and is also being shown to be useful in tracking parasitic diseases. Dracunculiasis, also known as Guinea worm disease, is caused by the Guinea worm, a nematode with a curious life cycle. The larvae of the parasite are found in freshwater, such as lakes, where they

Fig. 5.4 (a) The life cycle of the guinea worm. (b) Removing the worm from under the skin is currently one of the only ways to treat this horrible parasitic disease.

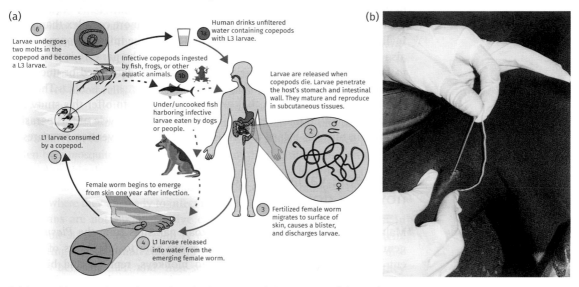

(a)

6 Larvae undergoes two molts in the copepod and becomes a L3 larvae.

Infective copepods ingested by fish, frogs, or other aquatic animals. 1b

1a Human drinks unfiltered water containing copepods with L3 larvae.

L1 larvae consumed by a copepod. 5

Under/uncooked fish harboring infective larvae eaten by dogs or people.

Larvae are released when copepods die. Larvae penetrate the host's stomach and intestinal wall. They mature and reproduce in subcutaneous tissues.

2

Female worm begins to emerge from skin one year after infection.

4 L1 larvae released into water from the emerging female worm.

3 Fertilized female worm migrates to surface of skin, causes a blister, and discharges larvae.

(b)

(a) https://www.cdc.gov/parasites/guineaworm/biology.html (b) CDC/The Carter Center.

are eaten by copepods (very small crustaceans). The parasite is transmitted to people when they drink or ingest contaminated water (see Fig 5.4). The larva, once inside a person, burrows through the stomach wall and develops into the adult worm in the subcutaneous tissue. This is where it gets really interesting. The female worm grows very large—up to 80 cm long, whilst the male worms are so tiny that very little is known about them, even how long they live. The females live just under the skin, usually on the leg, and over the course of a year or so, a very painful, sore, blister develops on the leg of the infected person. The pain is so intense that the best relief is to bathe the wound in water . . . and at this point, the adult female springs into action, and releases all of the larvae developing inside her, back into the water, thus completing the life cycle when they are eaten by copepods.

The Guinea Worm Eradication Programme (GWEP) has been under way since 1981. Along with other measures, such as health education, this has primarily involved filtering the drinking water. During the course of the GWEP, nearly 200 countries have been certified free of dracunculiasis, and fewer than ten countries remain to be certified free. Overall, cases have gone from over 30,000 in 1986 to just nineteen in 2017. However, since 2010 the number of cases of GWD in Chad have been increasing after an absence of any cases for a decade; the cause of this increase was puzzling and along with other investigative measures, the GWEP has turned to genomics to help unravel the mystery.

Along with an increase in the number of people infected, a large number of dogs were also infected. The genomes of the worms that were collected from infected people and dogs were sequenced. By comparing the genomes of these worms, we can infer, or predict, the familial relationship

attack every red-coat-wearing red blood cell. At some point, one parasite will switch from wearing a red coat to a blue coat: this parasite is now effectively camouflaged. The red-coat-wearing parasites will be killed, whilst the blue-coat wearers will survive and reproduce, until eventually most of the population is wearing blue coats. At this point, the immune system will catch up again and start recognizing blue coats. Not to be beaten, one parasite will switch again to a green coat, and the arms race will continue.

We only started to understand the molecular mechanism underlying this strategy in detail when the *Plasmodium* genome was sequenced and annotated. A substantial proportion of the *Plasmodium* genome encodes multigene families; these are genes that exist in multiple copies in the genome, with the copies being very similar (but not identical) to one another. They encode very similar proteins. When the genome was assembled it was noticed that these genes are located at the ends of the chromosomes, in what is known as the subtelomeric region. One of these multigene families is called the *var* family; genes from this family encode PfEMP1 proteins.

The *var* family has about sixty members, but interestingly these genes are being constantly shuffled by a process called genetic recombination, allowing many more than sixty *var* gene variants to be produced. Of course, the parasite has to make clever use of this vast amount of variation by only expressing one *var* gene at a time; if a parasite wore two different-coloured coats at the same time, this would halve the length of time taken to work through the whole repertoire of coats, halving the length of time the parasite can keep hiding from the immune system. The mechanisms by which *Plasmodium* parasites achieve this single-gene expression are still being actively investigated.

We learn more about antigenic variation—this time in the context of trypanosomes—in Case study 5.1.

Case study 5.1
Antigenic variation in trypanosomes

Malarial parasites are not the only organism that can 'change its coat'. Another organism which uses antigenic variation to evade destruction by the immune system is the African trypanosome, *Trypanosoma brucei* (see Fig A).

Different species of the *Trypanosoma* genus are able to infect either humans or animals. In humans, they cause a disease called African trypanosomiasis (also known as sleeping sickness). Trypanosome infections in livestock have a huge and devastating effect on African economies. Similar to *Plasmodium*, trypanosomes are transmitted by an insect vector: in this case, the tsetse fly (*Glossina* genus; Fig B). When an infected tsetse fly bites a human or animal host, the parasites are injected and go on to live extracellularly in the blood and interstitial spaces (ie in between the cells) of several tissues (brain, skin, and fat-storing tissues known as adipose). They also employ antigenic variation to evade the mammalian immune system (Fig C).

In the blood, trypanosomes cover over 95 per cent of their surface with a protein called variant surface glycoprotein (VSG for short). This protein effectively forms a shield around the parasite, protecting other, invariable, surface proteins from being exposed to the immune system. Much like the *Plasmodium* story, trypanosomes only express one VSG at a time, and they cycle through variants to keep ahead of the mammalian immune system.

The mechanics of trypanosome antigenic variation are slightly different to those of *Plasmodium* parasites. The genome of *T. brucei* has more than 1,000 VSG-encoding genes. Interestingly, about 80 per cent of these genes are actually incomplete: if they were to be transcribed and translated as they are, they would not produce a functional protein. What's more, there are only fifteen regions of the genome from which VSGs are actually transcribed. These regions are called bloodstream expression sites, or BESs, and they are located in the telomeric regions of the chromosomes. Only one BES is ever active

Fig. A The trypanosome parasites that cause the symptoms of sleeping sickness can clearly be seen here as the blue organisms amongst the red blood cells

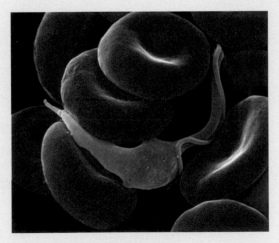

Clouds Hill Imaging Ltd/Science Photo Library

Fig. B Trypanosomes that cause sleeping sickness are transmitted by the tsetse fly

Image Quest Marine/Alamy Stock Photo

Fig. C The life cycle of the trypanosome parasite that causes sleeping sickness

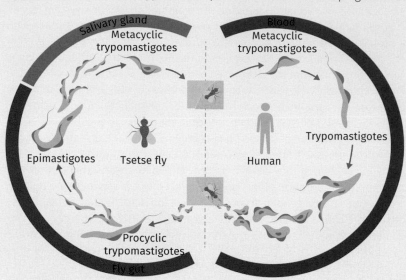

at a time. Much like *Plasmodium*, trypanosomes regulate VSG switching by both switching the BES that is active and rearranging the genome itself, via recombination, to produce new, functional VSGs.

Using genomics to control the spread of parasitic diseases

Not only has genomics taught us a lot about how parasites survive and thrive, but it also offers us a powerful tool to help fight the spread of parasitic diseases. A great recent example of this in action is the use of genetic engineering to manipulate the mosquito genome, contributing to malaria control.

Mosquitoes depend on blood meals from mammals for their survival (Fig 5.6), meaning that not only do they transmit parasites, but some would also consider them to be parasites themselves. Currently, the most common way to control mosquito populations is through the use of chemical insecticides. Whilst these are effective at killing mosquitoes, they also cause harm to other insects, potentially perturbing the ecosystem in unexpected ways. Mosquitoes are also increasingly evolving resistance to insecticides, reducing the efficacy of the limited number of chemical formulations that we have.

Fig. 5.6 The *Anopheles* mosquito—a parasite that introduces other parasites into the blood stream of its host

iStock.com/anatchant

Since the sequencing of the mosquito genome in the early 2000s, a much deeper, molecular understanding of mosquito biology has emerged. Genes and pathways have been characterized that are involved in mosquito viability, fertility, and their ability to transmit disease. This understanding has led scientists to speculate: could mosquitoes be genetically modified to make them resistant to infection by *Plasmodium* parasites? Or so that they die before they are able to transmit parasites to the next person? Or so that they cannot reproduce?

It has been frequently demonstrated that it is possible to introduce these traits to individual mosquitoes, but the difficulty comes in finding a way to increase the frequency of the trait in the population to such an extent that it has an effect on disease transmission. Releasing a handful of resistant mosquitoes into a wild population is not going to have an impact unless those mosquitoes can spread their resistant genes around the whole population. As might be expected, mosquitoes that have been genetically engineered are more often than not less fit than their wild-type siblings. Thus, we need to find a way to artificially drive these useful traits through populations.

Gene drive is the process whereby the inheritance of an allele is biased above what would be predicted by Mendelian genetics, causing that allele to spread through a population, even if it does not confer any fitness benefits. Genetic elements that drive themselves through populations are naturally occurring and found in all types of microbial life, but scientists boldly decided to see if they could hijack this genetic strategy to control the spread of vector-borne disease.

Essential genes are genes that an organism needs at least one copy of, without which it is unable to survive. When an essential gene is disrupted, it no longer produces a functional protein. Like all **diploid** organisms, mosquitoes have two copies of most of their genes, one maternal (ie, inherited from their mother) and one paternal (ie, inherited from their father). In some cases, if only one of those copies is disrupted the mosquito can make enough protein from the other copy of the gene to survive (such a gene is known as a recessive lethal). Mendelian genetics tell us that a diploid organism that has one functioning copy of a gene and one disrupted copy will pass on the functioning copy to 50 per cent of its offspring and the disrupted gene to the other 50 per cent. However, if the mosquito has got a perfect, 100 per cent efficient gene drive system in its germline (the cells that produce the sperm or eggs), this ratio will be skewed so that 100 per cent of its offspring inherit the disrupted gene. Essentially, a heterozygous individual will pass the gene drive system such as HEG (see Scientific approach 5.1) on to 100 per cent of its offspring as though it were homozygous, whilst remaining heterozygous in the rest of its body!

Try to imagine what would happen if a mosquito with one copy of an essential gene disrupted by a gene drive system were to be released into a wild population. That mosquito would survive perfectly well, because in most of the cells of its body it has another copy of the essential gene. But if the gene drive system is active in its germline, all of its gametes will inherit a copy of the disrupted version of the essential gene. When this mosquito mates with a wild mosquito, their offspring will also have one copy of the disrupted gene, but will survive fine with their other copy. The same thing will happen in *their* germline, so when they then mate with wild mosquitoes, all of their offspring will have one disrupted copy of the gene and one functioning copy. In this way, the disrupted gene will silently drive itself through the population. Eventually, it will reach such a frequency that every mosquito in the population has one copy. When these mosquitoes meet and mate, they will produce offspring that are homozygous for the disrupted gene; these mosquitoes will be not produce any functional protein, so they will die. The clever thing is that by the time mosquitoes start dying, the disrupted gene will already have spread throughout the population, so the fact that the disruption of this gene has such a severe effect on fitness hasn't stopped it from spreading. This strategy has been demonstrated in controlled mosquito populations in laboratory conditions, and is in the process of being developed for wild populations.

Scientific approach 5.1
An example of a gene drive system

Gene drives are important, as you have seen. How does gene drive actually work, at a molecular level? There are several naturally occurring gene drive systems, such as homing endonuclease genes (HEGs for short). HEGs are genes that encode proteins called nucleases. Nucleases are enzymes that cleave nucleotide chains, such as DNA. There are many different types of nucleases that have different **recognition sequences** (this is the specific sequence of DNA that the enzyme cuts). Cleavage of genomic DNA is normally a bad thing for a cell, and cells have evolved machinery to fix this when it happens. One of the processes they use to do this is called **homology-directed repair**. This is where the cell uses another very similar (**homologous**) piece of DNA as a template to help it repair the cut. The most similar piece of DNA present in the cell will normally be the homologous chromosome. If a diploid organism contains one copy of an HEG in its genome, and the HEG is active, the nuclease encoded by the HEG will make a cut in its target sequence in the homologous chromosome. When the cell repairs the break, it will use the region flanking the HEG itself (to which it is homologous) for repair (see Fig A). The HEG itself will also be copied to the

Fig. A Homing gene drive can be used as here to kill organisms that cause severe human or livestock disease—but only after the required genes have been driven into the majority of the population. Here, a mosquito has an active HEG in its germline. All gametes that it produces will have a disrupted copy of the gene. When this mosquito mates with another mosquito with an active HEG, their offspring will be homozygous for the disrupted version of the gene and they will die.

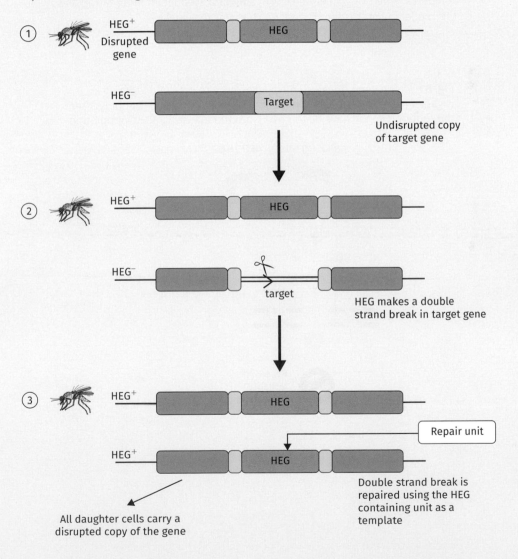

broken chromosome as part of the repair process. So, a cell that starts with one copy of the HEG will be converted to a cell with two copies, one on each homologous chromosome.

We can make this useful by engineering HEGs so that they target sequences within essential genes. This means that when the HEG cuts the homologous chromosome and copies itself across as part of the repair process, it also disrupts the essential gene. Until recently, engineering nucleases to target specific sequences was the bottleneck in this process, being extremely technically challenging. Now, we can use systems based on CRISPR (see chapter 7) to make targeted cuts in DNA much more easily (Fig B).

It can be difficult to imagine how gene drive works in a population. Fig C compares gene drive inheritance with standard Mendelian inheritance, and this modelling makes the process much easier to visualize.

Fig. B Homing gene drive can be used as here to produce lethality in organisms that cause severe human or livestock disease—but only after the required genes have been driven into the majority of the population

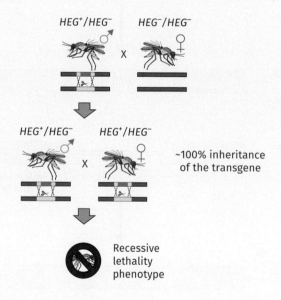

Fig. C In mosquitoes, as for any other sexually reproducing organism, heterozygous genetic elements have a 50 per cent chance of being inherited by the progeny and therefore its frequency remains constant in the population, or more likely, they will be gradually lost if they carry a fitness cost. A gene drive results in most or all progeny of heterozygotes receiving the driving genetic element; this allows the modification to spread rapidly throughout the population over a few generations (b).

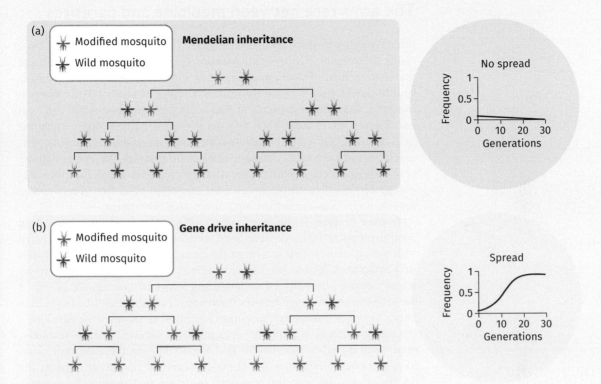

But is it ethical . . . ?

To many people, scientists or not, work like this is of incredible potential benefit. The main and most effective way of preventing malaria is still to keep away from mosquitoes. If the mosquitoes can be removed in a permanent way from malarial areas, surely that can only be a good thing. To the millions of people affected by malaria around the world, watching their children die or unable to work and support their families because of this terrible disease, the thought that one day the curse of malaria will be lifted by gene-modifying techniques must seem an impossible dream.

But science, even science with the best of intentions, does not exist in a vacuum. Some people have grave reservations about germline manipulation of the genome, as is used in these mosquitoes, or indeed about any kind of genetic manipulation at all. Some people are concerned about unexpected environmental effects from releasing genetically modified mosquitoes, whilst others worry that wiping out a whole species could have unpredictable effects on delicate ecosystems. There are also societal problems to be overcome. Imagine explaining to a remote community, with no genetics education, that you're going to reduce disease transmission by (initially, at least) releasing *more* mosquitoes into their local environment. That is a difficult sell!

The arms race between medicine and parasites

Effective anti-parasitic vaccines have proved very difficult to develop. As a result, there are no effective anti-parasite vaccines currently available, and the main method of treating infections is with drugs that kill the parasite. Unfortunately, only a relatively limited number of anti-parasitic drugs is available, and while it is excellent that we in the developed world can treat our children successfully when they get threadworms, that doesn't help the millions of people globally affected by much more serious parasitic diseases. As you have seen, parasites often have complex life cycles and are very difficult to target with drugs. And the countries which have the biggest problems with parasitic diseases are also often poor. This in turn means they do not necessarily get the biggest research investment.

Another problem is the development of parasite resistance. Mass drug administration (MDA) is the term used when whole communities are treated with a drug, without first screening the communities to see who is and who isn't infected. MDA has been used to effectively treat many communities where parasite infections are endemic and it has, without doubt, benefited the people who have been treated. However, repeated use of drugs over time puts a strong selection pressure onto parasites; if one parasite develops a chance mutation in its genome that enables it to survive the drug treatment, then when all other parasites in that population die (because they don't have this chance mutation and therefore 'survival advantage' against the drug), the parasite with the beneficial mutation will survive. Researchers are searching to understand how the drugs work, what they work against (what it is that the drug targets within the parasite), and why they are no longer effective at clearing the infection. Drug resistance has been reported with drugs against malaria, leishmania, trypanosomes, and helminths and poses a serious threat to efforts to the control of parasitic diseases.

We explore an example of a battle against drug resistance that researchers are fighting in Case study 5.2.

Case study 5.2

Fighting resistance to oxamniquine in blood fluke: Finding the needle in a haystack

Oxamniquine is one of just three drugs used to treat intestinal schistosomiasis, also known as bilharzia, caused by the parasitic blood fluke, *Schistosoma mansoni* (Fig A). It is only effective against *S. mansoni* and does not work against other human schistosomes (*S. japonicum* in Asia and *S. haematobium*

Fig. A Scanning EM of *Schistosoma mansoni,* the tiny worm which causes intestinal schistosomiasis

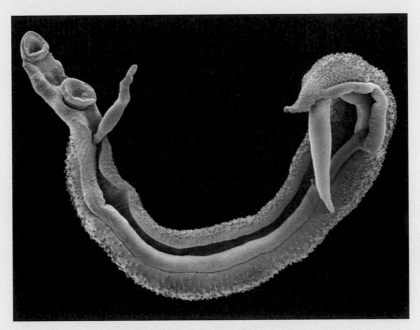

The Natural History Museum/Alamy Stock Photo

in Africa and Asia). Oxamniquine has been used successfully for MDA in South America but resistance emerged in Brazil in the 1970s.

The genome sequence and genetic map of *S. mansoni* have been used by researchers to investigate oxamniquine resistance in a beautifully elegant series of experiments. One interesting fact (amongst many!) about *Schistosoma* species is that the larval form of the parasite, called the sporocyst, is found in aquatic snails which act as intermediate hosts. Within the snail, the sporocyst replicates asexually, meaning that many genetically identical parasites are produced. This can be exploited to great effect in the lab. By infecting the snail with one, single *S. mansoni* larva, the offspring of that one parasite are all genetically identical. This allows the scientists to link the phenotype of the parasites with their genotype, because there are many identical parasites that can be examined in different ways for multiple purposes. Having the ability to effectively explore the relationship between a phenotype and a genotype is an extremely useful and powerful tool in the lab.

Researchers took the resistant strain and the sensitive strain of *S. mansoni* and performed a genetic cross (Fig B). The resulting offspring (the F1 generation) from that genetic cross were crossed again (ie mated with each other), producing the F2 generation. Adult parasites from the F1 and F2 generation were exposed to oxamniquine to see whether they were resistant or sensitive. The F1 generation was sensitive to oxamniquine. Of the F2 generation, 75 per cent were sensitive to the drug, and 25 per cent were resistant. This suggests that resistance to oxamniquine is controlled by a single gene recessive trait, according to Mendelian inheritance laws—exactly as demonstrated by Mendel's famous smooth and wrinkled pea experiments.

Fig. B The inheritance of resistance in *S. mansoni* follows a very predictable Mendelian pattern

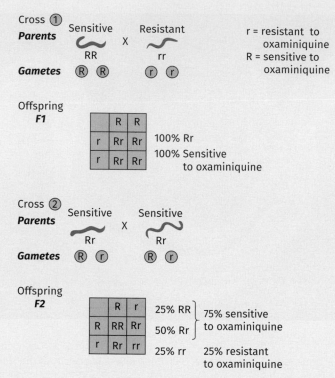

Using the clonal material available from the snails infected from single parasites (ie possessing parasites that all had identical genomes), the researchers determined the genotype of the parents, the F1 and the F2 generations using information from a genetic map of **microsatellite markers** that was previously constructed for *S. mansoni*. They found a specific region in the genome, at the end of one chromosome (chromosome 6), that was linked with resistance to oxamniquine, as determined by the pattern of inheritance of that region of the genome. The genes present in this single, **Quantitative Trait Locus (QTL)** could then be considered as potential targets of the drug oxamniquine. By examining the genes that occur in the genome at that location, it is possible to start to track which gene or genes might be responsible for resistance. Initially the QTL region contained 184 genes, which is quite a lot of genes to fully explore as candidates. In order to narrow down the search, the full genomes of the parents and two of the F1 offspring were sequenced. This sequence data meant that the inheritance of every single base pair of the genomes could be compared with each other and single nucleotide polymorphisms (SNPs) could be identified. Where SNPs were present between the resistant and sensitive strains, it was possible to further still narrow down the size of the QTL window to one which contained just sixteen genes. So, the number of needles in the genetic haystack had been reduced massively to sixteen—but still needed to be pruned further to find the culprit gene.

In both the sensitive and resistant parents, the same genes were present in the QTL region; this told the scientists that the resistance could not be due to a gene being completely absent as a result of the deletion of either the entire gene or an exon. The researchers established three criteria that they believed the gene or genes responsible for resistance to oxamniquine—known as OXA resistance—would possess (Fig C).

1. The gene or genes would contain base differences between the resistant and the sensitive parasites that would cause a change in the amino acids encoded by that gene. Remember, some nucleotide mutations (**synonymous mutations**) encode for the same amino acid, and therefore produce no change in the resulting protein, due to redundancy in the genetic code. Other mutations *do* lead to changes in the resulting amino acid and protein and are known as **non-synonymous** mutations.

2. Messenger RNA (mRNA) transcripts from the gene or genes would have to be present in the adult parasites if they are involved in the mechanism of resistance—remember that not all genes are expressed at every stage of the life cycle of an organism.

3. The size of the predicted protein could be informative and so should be taken into account when narrowing down the list. Some of the proteins that are expressed by *S. mansoni* adults have already been studied, and so some information on them, such as their size and some properties, are already known.

Fig. C Three criteria were used to narrow down potential candidate genes: (1) non-synonymous SNPs differing between the parent parasite; (2) gene expression in the adult worm; and (3) a protein size between 25 and 35 kDA. Just one gene met all three criteria: Smp_089320.

Gene	Notes	Has non-synonymous SNPs differing between parent parasites	Detectable expression in adult worms (RNA seq)	Protein size between 25 and 35 kDa
Smp_162320	Hypothetical protein	Yes	Yes	No
Smp_074430	Hypothetical protein	No	Yes	No
Smp_179290	ATPase family	Yes	Yes	No
Smp_179270	Heterogenous nuclear ribonucleo protein	No	Yes	No
Smp_179260	Alpha galactosidase	No	Yes	No
Smp_179250	Alpha galactosidase	No	Yes	No
Smp_186710	Alpha galactosidase	Yes	No	No
Smp_170830	Alpha galactosidase	No	No	Yes
Smp_089270	Alpha galactosidase	No	Yes	Yes
Smp_170840	Alpha galactosidase	No	Yes	No
Smp_170850	Alpha galactosidase	No	No	No
Smp_089290	Alpha galactosidase	No	Yes	No
Smp_170870	PDZ domain containing protein GIPC3	No	Yes	No
Smp_089320	Cell wall integrity and stress response	Yes	Yes	Yes
Smp_089330	NAD-dependent epimerase: dehydratase	No	No	No
Smp_192140	NAD-dependent epimerase: dehydratase	No	No	No
Smp_119060	NAD-dependent epimerase: dehydratase	No	No	Yes

It is important to note at this point that oxamniquine is a pro-drug. It has to be activated by the enzyme sulfotransferase before it becomes active in the body of the worms. Previous experiments had shown scientists that the parasite gene responsible for activating oxamniquine had properties of an enzyme known as a sulfotransferase, and that three of the sixteen genes showed homology (sequence similarity) to sulfotransferases. Now, based on the above criteria, seven of our sixteen candidate genes fulfilled the first criterion and possessed non-synonymous changes between the sensitive and resistant strains, six genes were excluded based on the second criterion because they were not expressed at all, and four of the sixteen genes had predicted proteins that were the same mass as the protein that had been shown in laboratory experiments to be responsible for activating oxamniquine. Just one gene, memorably called Smp_089320, met all three criteria, and it had homology to sulfotransferases. So, a promising lead . . .

In the laboratory, the researchers then set about validating this candidate gene. They included in their validation tests some of the other genes on the shortlist which fulfilled at least two of the criteria set, so as not to put all their eggs in one basket. First, they produced recombinant proteins from the candidate genes in the oxamniquine sensitive parents, and tested whether these proteins activated oxamniquine in resistant worms. The theory behind this is that resistant worms cannot normally activate oxamniquine, but if oxamniquine is activated in the presence of a protein product of one of the candidate genes, then it suggests strongly that that gene is responsible for, or involved in, activating oxamniquine. The protein from the main suspect gene, Smp_089320, was the only one which activated oxamniquine in resistant worm extracts.

To further test this, **RNA interference** (RNAi) was used to disrupt Smp_089320. RNAi is an endogenous cellular process that is involved in antiviral immunity; in experiments such as these, scientists can hijack this system to inhibit the expression of genes that they are interested in. Expression of the gene was reduced almost completely by RNAi against it, and in normally sensitive parasites, knocking down their version of Smp_089320 meant that they became resistant; oxamniquine could no longer be activated and the parasite survived.

Close examination of the sequence of the genes in sensitive and resistant parasites revealed that they differed in just two positions, and researchers were able to determine which mutation was the one responsible for resistance. Once they had pinpointed the location of the mutation responsible for resistance in these laboratory-bred parasites, they examined the genomes of *S. mansoni* parasites obtained from the field. They discovered that the same mutations in Smp_089320 were also present in resistant parasites from natural, disease-causing populations in the field.

Furthermore, exploring the three-dimensional crystal structure of the product of Smp_089320 allowed the researchers to understand how this protein differs in sensitive and resistant parasites, and which parts of the protein actually come into contact with the oxamniquine molecule. When genes in related Schistosome species, *S. japonicum* and *S. haematobium* were examined (remember that both are completely resistant to oxamniquine), it was found that mutations were present in the key areas of Smp_089320 homologues that prevented binding to oxamniquine, revealing the underlying reason behind oxamniquine only being effective against *S. mansoni*, and not *S. japonicum* or *S. haematobium*.

   Thus, the genomic sequence of *S. mansoni* helped researchers to unravel the story beneath emerging drug resistance and is a crucial starting block for much-needed further research into alternative treatments for this parasite.

Keeping parasites in the lab

You might be wondering how we manage to study parasites that infect humans. You might also be thinking that we surely don't infect people deliberately in order to study the parasites (you'd be right!). If we want to study parasites so that we can help to reduce the impact they have on people, then how can we do this? Parasites are a great problem to agriculture, livestock, and our pets, as well as to humans directly, and so it is important that we study parasites of animals too—what's more, animal parasites can be easier to get hold of than human parasites. Additionally, and crucially, there are many similarities between animal parasites and human parasites, and so what is learnt about one can often be extrapolated to the other.

   Parasites that infect animals and humans can be studied by obtaining 'field' or 'clinical' samples. For example, sheep poo in pastures can contain larvae from several parasitic nematodes, including, for example, the barbers pole worm (*Haemonchus contortus*). Similarly, pig, cow, horse, dog, or cat poo (amongst others) can contain many different parasites (Fig 5.7). Poo, therefore, can be a good source of parasite material as it is an important part of the parasite life cycle—the parasite is spread from host to host through faeces.

   Field parasites samples can also be obtained from people, and depending on where the parasite resides in the body of the host, different techniques

Fig. 5.7 Pigs, pets, and poo—all sorts of animals, and their waste, can be used as a source of parasites for the laboratory.

© Anthony Short

are used. Leishmania can be taken from skin lesion swabs with either a biopsy or a much less invasive approach using tape strips. Parasites that reside somewhere in the gastrointestinal or urinogenital tracts of humans can be obtained from the urine or faeces; some flukes (eg schistosomes) and gastrointestinal nematodes (eg whipworms) can be directly sampled from human faeces. One technique, called Kato Katz, involves smearing the faecal sample onto a glass slide, which is then examined under a microscope to search for parasites that can then be used for diagnostic purposes and for research experiments. Other methods include flotation of the faeces and subsequent separation of the parasite from the faecal material, and blood or biopsy samples, although these can be quite invasive and unpleasant for the patient.

Particular life-cycle stages of some parasites are quite inaccessible because the types of cells or tissue in which the parasite resides are in parts of the body that are not easy to access. Hookworms (such as *Necator americanus* and *Ancylostoma duodenale*) and threadworms (such as *Strongyloides stercoralis*) infect people when the larvae burrow or penetrate through the skin. Once inside the body, the larvae are carried around the body in the bloodstream, ending up in the lungs, where they are eventually coughed up and swallowed into the intestine where they develop into adults. As a result, we really can't get hold of the larvae or the adults from these species. When people are treated with **anthelmintics** (medicines that kill helminths), in the case of hookworms and threadworms, the parasites are expelled in the faeces, from which they can be collected, and this is commonly used in studies that incorporate MDA.

As you have read in this chapter so far, scientists often want to test certain conditions or to manipulate the parasite genetically in order to answer specific questions, and in these instances, it is necessary to maintain the parasite life cycle in the laboratory. Some species are relatively easy to maintain in the laboratory. The mouse whipworm, *Trichuris muris*, has a relatively straightforward life cycle that only involves the mouse host. In comparison, the blood fluke, *Schistosoma mansoni*, has a complex, indirect life cycle that involves not only the mammalian host (humans in nature, rodents in the laboratory) but also a particular type of aquatic snail. Thus, in order to keep *S. mansoni* parasites in the lab, colonies of aquatic snails need to be kept too. The human malaria parasite, *Plasmodium falciparum*, can be cultured together with human blood in flasks in the lab.

It is always of paramount importance that laboratory animals are used only where there are no other options to perform the same experiment, and indeed, much research is going on to find replacements, such as using things called organoids (see Scientific approach 5.2 on the 3R's and organoids.) However, the advantages of maintaining a parasite in the laboratory include being able to directly study the human-infective form, if that parasite is also able to infect a different species, such as mice. Many parallels can be drawn from such studies on laboratory animals, but researchers are always looking to find approaches that avoid the use of animals.

Scientific approach 5.2
Animal alternatives: the 3Rs and organoids

The 3Rs of animal research are Replacement, Reduction, and Refinement. Before any experiment is conducted using animals, scientists must first consider whether there are alternative methods to **R**eplace the use of animals, how to **R**educe the number of animals used per experiment, and how to **R**efine the methods used to reduce animal suffering.

One exciting new technology that could significantly reduce the number of animals used in research is based on the use of structures called organoids (Fig A). These are miniature versions of organs, which contain all of the different cell types that make up an organ. Each organoid is grown *in vitro* from a small number of cells taken from an organ. This means that a single mouse can be used to produce many organoids, hugely increasing the number of experiments that can be performed. Even better, organoids can be produced from human cells. It's not all plain sailing though; huge amounts of work are needed to work out the exact culture conditions required for organoids to grow and differentiate properly. What's more, while organoids may be useful for studying how parasites interact with specific organs, animals will likely continue to be needed to maintain parasites with complex, multi-organ life cycles.

Fig A Three-dimensional organoids such as these intestinal organoids grown *in vitro* have the potential to replace many animal experiments in future

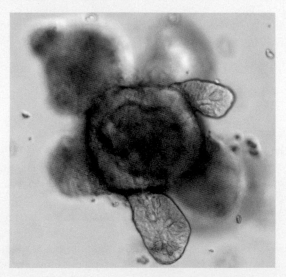

St Johnston, D (2015). 'The Renaissance of Developmental Biology.' *PLoS Biol* 13(5) e1002149. https://doi.org/10.1371/journal.pbio.1002149/Wikimedia Commons/CC BY 4.0

The field of parasite genomics has emerged since the last decade of the twentieth century and continues to gain pace. The genomics groundwork has proven to be an essential tool for researchers, not only enabling people to understand some of the interesting quirks in parasite genomes themselves, but also revealing crucial chinks in parasite armoury that can potentially be exploited to produce effective treatments. As technology continues to advance, it is hoped that some of the most widespread infectious diseases in the world, which disproportionately affect the most disadvantaged people, can begin to be tackled head on.

Chapter summary

- Parasites and their genomes are hugely diverse.
- Genomics can be used for inferring relatedness, helping us track how parasites are transmitted.
- Some protozoan parasites, such as malaria and trypanosomes, use antigenic variation to evade destruction by the mammalian immune system.
- Genetic engineering can be used to manipulate the genomes of parasites and their vectors, and is a novel way to tackle disease transmission.
- Genomics can give major insights into the mechanisms behind drug resistance in parasites.
- A limiting factor in parasite genomics is often the availability of parasite material: parasites can be small and live in places that are hard to access.
- Animal hosts and specialized environments are often needed to keep and study parasites in the lab.

Further reading

1. How genomics is being used to defeat neglected tropical diseases: https://www.yourgenome.org/stories/how-is-genomics-being-used-to-tackle-neglected-tropical-diseases.
2. The ongoing battle against drug-resistant malaria: https://www.yourgenome.org/stories/the-ongoing-battle-against-drug-resistant-malaria
or https://www.yourgenome.org/stories/malaria-the-master-of-disguise or
https://www.yourgenome.org/stories/a-weapon-against-malaria.
3. More on gene drive and the promise of wiping out mosquitoes: https://www.bbc.co.uk/news/science-environment-45628905 or
https://www.economist.com/briefing/2018/11/08/the-promise-and-peril-of-gene-drives.

Discussion questions

5.1 Discuss the benefits and possible problems of using gene drive to wipe out disease-carrying mosquitoes.

5.2 Explain why it is useful to understand the genetic basis of drug resistance.

5.3 What are the pros and cons of using organoids to study host–parasite interactions?

5.4 Should we try and wipe out all human parasites? Research and discuss any advantages to humans in living with parasites.

6 HUMAN EVOLUTION

Exploring human evolution may make you think of fossil skulls glaring out from your screen, or bones being excavated from the depths of a remote cave, or perhaps monitoring chimpanzees in a tropical rainforest—but you may be surprised to discover that genome sequences are now central to our study of human evolution. In this chapter, we will draw on information from all of these sources, and others, because we need them all to build the fullest picture that we can of how humans evolved, but we will give a special emphasis to the information that comes from genomes.

Since the time of Darwin, we have understood that we are one of the great apes, sharing an ancestor with them millions of years ago. However, our domination of the planet today contrasts starkly with their critically endangered status. When, how, and why did these different paths emerge? Why are we so abundant and so widespread? Why is there just one species of human compared with the variety of species of all the other great apes? We will see that the small populations of multiple species and subspecies seen in the other great apes were also found in humans until recently—that is, recently on an evolutionary timescale, so around 40,000–50,000 years ago!

Some key changes set us on a different path, and we still don't know what most of these changes were—they remain one of the big mysteries in the field. But genome sequences tell us how we differ genetically from other apes, fossils fill in details of some of the extinct species, and **ancient DNA (aDNA)** can, in favourable circumstances, even allow us to read their genome sequences. Using sequences from diverse present-day people as well, we will see how a rare African ape expanded over the whole world, constantly split into different populations, and then mixed again with whoever they met without prejudice (including with other types of humans like **Neanderthals**). These amazing animals adapted to different environments, foods, and

Fig. 6.1 *Homo sapiens*—we have one of the lowest genetic diversities, but the greatest range and biggest population of all the great apes on Earth.

Rawpixel.com/Shutterstock.com

pathogens, and came to create art, music, writing, education, money, laws, and every aspect of our present lifestyle and society that you may love or hate. This is the story of an extraordinary species, of you and me, of **Homo sapiens** (see Fig 6.1).

Humans as great apes

In 1991, when zoos had cages rather than enclosures or habitats, Jared Diamond wrote his popular science book *The Rise and Fall of the Third Chimpanzee* and threw out the following challenge:

> The next time that you visit a zoo, make a point of walking past the ape cages. Imagine that the apes had lost most of their hair, and imagine a cage nearby holding some unfortunate people who had no clothes and couldn't speak but were otherwise normal. Now try guessing how similar these apes are to ourselves genetically.

He considered us, humans, as a third chimpanzee, the other two being the common chimpanzee and the bonobo.

According to studies of our anatomy, we humans are most closely related to chimpanzees, bonobos, gorillas, and orang-utans, among living species. Genetically, we are also very similar. Humans have forty-six chromosomes, while the other great apes have forty-eight, but this difference in number is just the consequence of the fusion of two ape chromosomes on the human line. We now know the genome sequences of all these species, and see that the sequence difference between humans and chimpanzees or bonobos is only around 1.4 per cent. For comparison, the sequence difference between two random humans is about 0.1 per cent. Humans and gorillas differ by around 1.8 per cent, and humans and orang-utans by about 3.4 per cent.

Using this information, together with comparisons between the other species, we can draw a family tree as shown in Fig 6.2. In addition, using measurements of the **mutation rate**, we can estimate the times when the species separated: 5–7 million years ago for humans and chimpanzees, 8–10 million years ago for humans and gorillas, and 12–16 million years ago for humans and orang-utans. To put these numbers into an evolutionary context, 5 million years is about one-thousandth of the age of the earth.

Although we humans are one of the great apes, and shared the same evolutionary history as our closest relatives until a few million years ago, we now differ from all the other great apes in several important ways. All the others have multiple species and subspecies: bonobos and Western, Nigerian, Central, and Eastern chimpanzees (two species, one of which has four subspecies); Western lowland, Cross river, Eastern lowland, and mountain gorillas (two species, each with two subspecies); and Sumatran, Bornean, and Tapanuli orang-utans (three species) (see Fig 6.2). But humans all belong to one species with no subspecies.

All of the other great apes live in very restricted areas of the tropics in Africa or Southeast Asia, and have population sizes ranging from a few hundred (mountain gorillas) to approximately 250,000 (Eastern chimpanzees) (see Figs 6.2 and 6.3). In contrast, humans live in almost every part of the globe, thriving in environments that can include deserts, the arctic ice cap, and mountain tops, and have a population size of over 7 billion: more than 20,000 times higher than the other great apes. We also stand out in many aspects of our appearance and behaviour; these differences are discussed below in the section 'What changes made us human?'

From the genome sequences, we can also measure the level of **genetic diversity** within each species or subspecies, by looking at the differences in the nucleotide sequences. The other great ape species or subspecies mostly have higher genetic diversity than we do, despite their tiny population sizes (see Fig 6.2). One way to think about the similarities and differences listed in this section is that humans are the equivalent of a great ape species where only a single subspecies with low genetic diversity survives, yet has expanded enormously in range and numbers. Studies of fossils, described in the next section, 'Human fossils and archaeology', show that there were indeed additional types of human in the past, and so this thinking is close to the truth.

Human fossils and archaeology

Fossils are the remains of living organisms, or their traces such as footprints, that have become preserved in stone. They can survive for millions of years and thus complement studies of living species. In particular, they can tell us about our ancestors in a direct way, and about species who lived in the past, but have no descendants now.

Fig. 6.4 Fossil skeletons of (a) *Australopithecus afarensis* (Lucy, 3.2 million years old, Ethiopia), (b) *Homo erectus* (Nariokotome, 1.6 million years old, Kenya), (c) *Homo neanderthalensis* and *Homo sapiens*

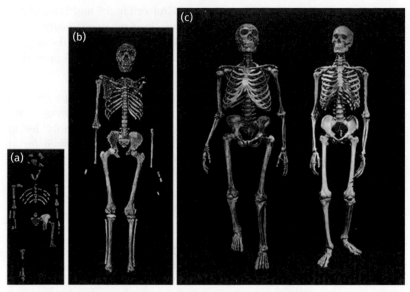

(a) Puwadol Jaturawutthichai/Alamy Stock Photo

Fig. 6.5 Neanderthals, Denisovans, and humans descend from a shared ancestral species who lived 500,000–600,000 years ago. There were several inter-breeding events (horizontal arrows) after the separation.

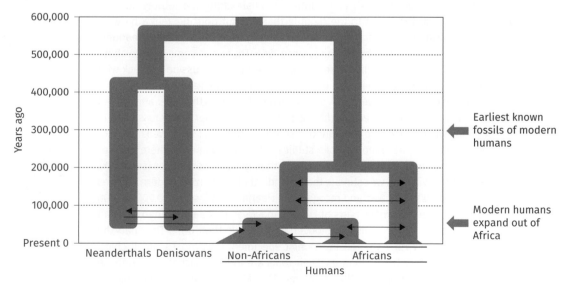

of humans, hobbits, and many other types in 'The Lord of the Rings' was perhaps closer to the truth 60,000 years ago than anyone imagined!

- Genome sequences of fossil humans, Neanderthals, and Denisovans have been determined from ancient DNA (see Fig 6.5 and Scientific approach 6.1), and reveal that these early humans occasionally inter-bred with one another, most strikingly illustrated by the discovery in Siberia of a fossil fragment from the teenage daughter of a Neanderthal mother and Denisovan father.

We know from fossil footprints that our ancestors walked upright well before 2 million years ago. Much more information about their lives comes from archaeology, the study of human activity from their tools, living places, and any junk they left behind. Stone tools have been used for at least 3.3 million years, fire for perhaps a million years, clothes for 70,000 years according to a genetic study of when body lice originated, art for about the same length of time, and farming for 10,000 years. So, the lifestyle we think of as typical of humans developed gradually.

Bigger picture 6.1

Who were the Neanderthals, and what happened to them?

Popular views of Neanderthals are so diverse that they have been summarized in the 'subway test'. This thought experiment asks how people would react if they saw a Neanderthal with smart clothes and hair sitting in a subway carriage. Would they sit down next to the Neanderthal without noticing, feel a bit worried and sit at a safe distance, or run screaming from the carriage? Neanderthals have been studied by scientists since 1864. There was some debate about whether the first fossils discovered were derived from a Russian Cossack who had lived a life of pain because of **rickets**, resulting in large brow ridges, or a distinct type of human! Eventually scientists decided it was the latter and named the new species *Homo neanderthalensis* after the Neander valley fossil site in Germany where they were first found. A century-and-a-half later, we know a lot about them from many more fossils, studies of the environments where their remains were found—and most recently from their genome sequences (see Fig A).

From DNA evidence, we estimate that Neanderthals split from modern humans around 500,000 years ago, and recognizable Neanderthal fossils 430,000–40,000 years old come from many parts of Europe and Western Asia. Neanderthals had a short, strong, stocky build and longer, lower skulls than ours. They had the prominent brow ridges noted in the first specimen, large noses, small chins, and slightly larger brains than us. Their tools are indistinguishable from those of modern humans living at the same time. They also used fire, gathered plants, hunted large animals, buried their dead, and probably created art. Yet by 40,000 years ago they completely disappear from the fossil record. Why?

The culprit is us: our ancestors, *Homo sapiens*, expanded from Africa into Neanderthal territories, reaching their last strongholds in Europe by 45,000 years ago. We don't know whether these early modern humans were more efficient at gathering food and occupying the best living sites than the Neanderthals and so outcompeted them, whether they spread diseases the Neanderthals had no resistance to, or whether they were just more successful in direct conflict. But the result was the rapid extinction of the Neanderthals . . .

Or perhaps not, depending on how you look at it. By comparing human and Neanderthal genome sequences, we see that all present-day non-Africans carry about 2 per cent of Neanderthal DNA. This is because humans migrating out of Africa met Neanderthals around 55,000 years ago, probably in the Middle East. Interbreeding took place and the resulting children were integrated into the *Homo sapiens* community. This Neanderthal DNA influences our hair and skin colour, sleeping pattern, virus resistance, diabetes risk, and many other characteristics to this day. For example, carriers of a region derived from Neanderthals on chromosome 2 near the *ASB1* gene are more likely to be 'evening people' who like to take naps during the day, and this region is more common in people from northern latitudes. The most bizarre association reported is between a Neanderthal region on chromosome 11 and a decreased tendency to sneeze after eating dark chocolate. What is the evolutionary importance of such an association? We have no idea: perhaps it is just chance, or perhaps the importance lies in some other effect linked to one of the ingredients or to the sneezing reaction . . .

Fig. A This reconstruction of a Neanderthal face based on DNA evidence as well as fossil remains brings these ancient hominids vividly to life

Richard Gray/Alamy Stock Photo

Scientific approach 6.1
Finding and analysing ancient DNA (aDNA)

If we want to know about the human genetic past, analysing DNA from ancient remains (ancient DNA or aDNA) seems the best and most direct approach. But in practice this is very challenging, in two particular ways. First, the process raises a considerable ethical question—is it acceptable for well-funded labs in developed countries to collect remains from anywhere in the world and work on them? Second, even when the collection and analysis of human material can be done with proper consent, it is still technically very demanding. Ancient DNA can be present in bones, hair, **coprolites** (fossil poo) and even in soil, but it is there only in tiny amounts. Worse, it is broken down into small fragments and damaged (chemically altered). And worst of all, it is contaminated, because DNA is everywhere: from bacteria and fungi in the surrounding soil to the skin cells shed continuously by all of us, including the scientists working on the aDNA. Scientists have developed several practices to get around these problems:

- Choose cold, dry environments like Siberia where the conditions ensure that the deterioration of the genetic material is kept to a minimum.

- Use bones such as particularly hard parts of the skull, which preserve DNA best.

Fig. A Working on ancient DNA and avoiding contamination

- Collect samples carefully without touching them directly to reduce the risk of contamination with modern human DNA.
- Extract DNA in a dedicated clean room—again, to minimize contamination (see Fig A).
- Work on a large scale or enrich the human DNA using special techniques developed for this purpose.

It is usually impossible to follow all of the guidelines. Nevertheless, aDNA is transforming our understanding of human evolution. We have genome sequences from Neanderthals and discovered that our ancestors inter-bred with them. We have identified a new type of ancient human, **Denisovans**, and discovered that they inter-bred with both Neanderthals and our ancestors. We also have data from thousands of early modern humans from many parts of the world, which is beginning to reveal the complexity of our past, including examples such as the blue eyes and dark skin of people in England 9,000 years ago.

What changes made us human?

We differ from all the other great apes in many ways, some obvious and others less so. These include physical characteristics such as our upright walking, larger brains, reduced body hair, **opposable** thumb and absence of a penis bone (see Fig 6.6). The differences also include social characteristics such as our larger group sizes, ability to speak and cry, and the potential for full artistic expression. In some cases, we know from fossils or archaeology when these characteristics appeared: upright walking evolved before 3.5 million years ago and artistic expression appeared by 70,000 years ago, for example (see 'Human fossils and archaeology' above). In other cases, such as speaking or crying, we have no idea when these behaviours appeared.

Often, genetic changes were necessary for these characteristics to appear. This does not mean that there is 'a gene for' any of them, or even a single mutation that made a big difference. More likely, multiple, small genetic changes contributed. How can we identify these?

When we compare our genome to that of a chimpanzee, the 1.4 per cent difference we noted earlier translates into around 35 million differences, half of which arose on the chimpanzee lineage and the other half on the human lineage. Of the 17–18 million differences on the human lineage, most have no detectable effect, but among them are the tiny fraction responsible for the changes mentioned above. How can we pick these out?

The answer at the moment is—there is no good way, so at present we do not know what most of the important genetic changes were! But this does not stop scientists from trying to find them, and we are making some progress. One approach is to start by identifying all the changes that are likely

Fig. 6.6 The penis bone of a macaque—humans, unlike monkeys and other apes, do not have one

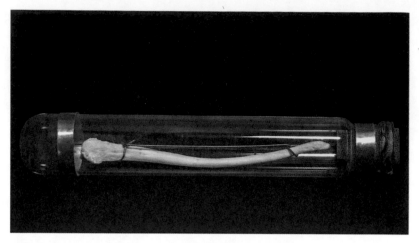

Didier Descouens/Muséum de Toulouse. CC-BY-SA

to have functional consequences. These may lie within functional elements such as genes, or within the regions that control the expression of genes. Alternatively, we can use our understanding of DNA evolution: important regions evolve slowly because many changes that may arise within them will be disadvantageous and therefore will not survive in the population, so the level of genetic variation in these regions is low. A second approach is to examine the sequences of species like Neanderthals. This way, we can determine whether a human-specific change arose before or after the human–Neanderthal split about 500,000 years ago, and thus whether it can be relevant to a particular human characteristic. Consequently, sequences which are evolving slowly in most species but show a difference between humans, chimpanzees, and Neanderthals, and lie in a functional region, are good candidates.

One gene that meets these criteria and has attracted much attention is *FOXP2*. It evolves slowly: there are only three differences in its **amino acid** sequence between humans and mice, yet two of these occurred on the human line after the human–chimpanzee split. It also has a very interesting function: people with a damaged copy have speech difficulties. Is it therefore a genetic contributor to human speech ability? One way to test this possibility is to engineer the two human-specific amino acid changes into mice. These mice do indeed show some subtle differences in the way they squeak: reasonable confirmation of its relevance to human speech. The two amino acid changes are shared with Neanderthals, so any contribution to speech would have been present in them as well. In addition, cats and a few other animals also have one of the two amino acid changes. While *FOXP2* evolution probably did contribute to human speech ability, understanding the details of its contribution is complex and remains under investigation.

But this is just one of many candidates. Others include *ASPM*, *BOLA2*, *CDK5RAP2*, *GADD45G*, *HAR1F*, *HYDIN*, *MCPH1*, *MYH16*, *NOTCH2NL*, *PAK2*, *PDE4DIP*, *SLC6A13*, and *SRGAP2*. One of these is discussed in Case study 6.1, but undoubtedly far more remain to be identified.

Case study 6.1
A gene that contributed to our big brains: *NOTCH2NL*

NOTCH2 is the name of a gene in humans and many other animals that is essential for their general development. Humans alone have, in addition to the regular gene, copies of a closely related gene called *NOTCH2NL*. Both the complex evolutionary history and the function of *NOTCH2NL* have been studied in some detail.

Around 14 million years ago, a duplication of part of *NOTCH2* (leaving the original *NOTCH2* intact) created a second but non-functional copy, which is therefore present now in gorillas and chimpanzees (who actually have several copies of it). But in human ancestors 3–4 million years ago, a mutation replaced the missing part of the gene. This then became functional again as *NOTCH2NL*, but still differed from the original *NOTCH2* gene. Soon afterwards, *NOTCH2NL* itself was duplicated twice, so humans (and Neanderthals) have three copies of this new functional gene, as you can see in Fig A.

NOTCH2NL is **expressed** (that is, used and performing its function) in some brain cells of the developing **foetus**. Here, it influences these cells to divide more times, with the consequences you might expect: the mature brain

Fig. A The complex evolutionary history of the human-specific *NOTCH2NL* gene

contains more cells and is therefore bigger. This is thus one part of the reason why our brains are bigger than those of chimpanzees or gorillas.

There is another ironic twist to this story. Mutations in the region containing the three copies of *NOTCH2NL* can lead to severe developmental disorders where the head is unusually large or small, or to milder conditions such as **autism**. The same changes that benefited humans during their evolutionary history also make us more susceptible to some disorders.

Our expansion out of Africa

Modern humans originated in Africa, 'the cradle of humankind', around 300,000 years ago and expanded into the rest of world around 55,000 years ago (see the section 'Human fossils and archaeology' above), so we now inhabit the entire land surface of the planet, apart from a few of the most hostile deserts, mountains, or polar ice caps (see Fig 6.3). It must have been a remarkable journey for our ancestors to colonize the entire globe in times when travelling was done on foot or by simple boat. Unfortunately, there is no logbook documenting how they did it. But we can reconstruct parts of the journey from the fossil and archaeological records, and increasingly from genetic data.

The world when modern humans left Africa looked somewhat different from today, as you can see in Fig. 6.7, which shows the world about 60,000 years ago with arrows showing migration times and points of potential interbreeding between early humans and other hominids. The temperature was colder and large ice sheets covered most of North America and northern Europe, making the sea level lower due to water trapped in ice. As a result, what are now the British Isles were connected to the European mainland. Similarly, one could have walked from Siberia to North America on dry land.

The route of the human expansion 'out of Africa' most likely followed an exit via the Middle East, and then spread into Asia and Europe, as you can see in Fig 6.7. Dating of archaeological sites across the world has enabled researchers to time the main migration events, and has led to some surprising findings. For instance, the oldest human sites from Australia are much older than any found in Europe! Although this might seem puzzling, considering how much further Australia is from Africa, people perhaps found following the warm coastline of South Asia to Australia more attractive than investigating the cold, barren route to Europe. But by 45,000 years ago, much of Asia, Europe, and Australia were populated. Expansion into the Americas occurred through **Beringia** (the region where Siberia meets Alaska) 15,000–20,000 years ago, and expansion into the remote islands scattered throughout the Pacific Ocean happened only in the past few thousand years.

Using genetic data from present-day people sampled from around the world, we find that their genetic diversity follows on average a simple geographical pattern, decreasing with increasing distance from East Africa (see Fig 6.8). This further supports an origin of modern humans in Africa, where the highest levels of genetic diversity lie, while the pattern in the rest of the

Fig. 6.7 This map shows possible routes of human migration out of Africa into the rest of the world, migrations which began over 60000 years ago (60kya). Possible locations of interbreeding with Neanderthals and Denisovans are marked with red and yellow stars, respectively. Note the map does not represent the world 60,000 years ago (60kya) but indicates possible migrations over a more modern map.

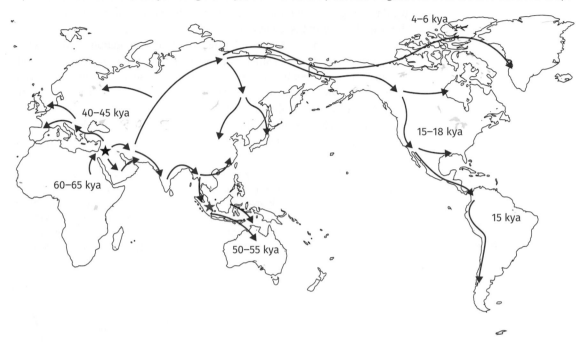

Fig. 6.8 Genetic diversity decreases with distance from East Africa.

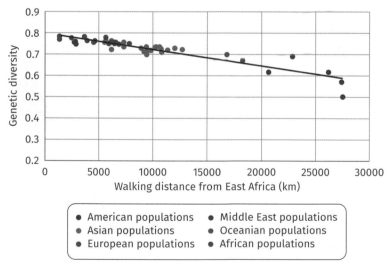

world can be explained by models where only a small subset of the African variation moved out because a relatively small population—perhaps a few thousand—took part in the migration. This temporary reduction in size is known as a bottleneck or founder event (see chapter 2). Further bottlenecks would, using the same logic, explain the decrease in diversity throughout the rest of the world (see Fig 6.9).

We saw earlier in Bigger picture 6.1 that the genomes of ancient Neanderthals have been sequenced, allowing their interbreeding with modern humans to be identified, and revealing the presence of about 2 per cent Neanderthal DNA in all present-day non-Africans. A single interbreeding event between the population emerging from Africa 55,000 years ago and the Neanderthals resident in the Near East at that time would explain the Neanderthal DNA in present-day populations. Other aDNA studies have led

Fig. 6.9 Population size changes over the last 300,000 years. Each line shows a different present-day population. Early on, the ancestors of all present-day populations were the same size, because they were the same population. Later, non-African populations (labelled B) began to diverge from Africans (labelled A.). The bottleneck associated with the migration out of Africa is clear. In recent times, some populations have expanded enormously.

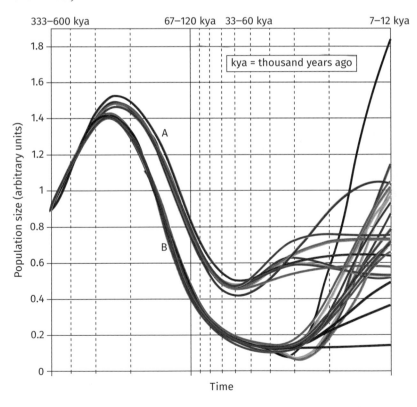

to an even more surprising discovery. A finger bone between 74,000 and 82,000 years old from Denisova Cave in Siberia was found to belong to a previously unknown and now extinct human branch related to Neanderthals, known as Denisovans, after the cave. We know almost nothing about what the Denisovans looked like, as no other remains apart from a few isolated teeth and a jaw bone have been found so far. But we do know that they also mixed with modern humans, most likely somewhere in South-East Asia around 50,000 years ago. As a result, present-day Asians carry trace amounts of Denisovan DNA, while Papuans and Aboriginal Australians carry around 3 per cent; perhaps there was more than one interbreeding event.

After the initial expansion into much of the world between 45,000 and 55,000 years ago, people did not stop moving around. From studies of aDNA, we can identify some of the later changes, and we are just beginning to understand the vast extent of the movements and other events: expansion of some populations, extinction of others, or merging of different populations. For example, aDNA from a four-year-old boy who lived in central Siberia 24,000 years ago showed that the Siberian population then was very different to the present-day people living in this region. But more interestingly, there were stronger similarities between the ancient boy's DNA and present-day Native Americans, and some lesser similarities to present-day Europeans. Native Americans and Europeans thus contain shared ancestry from ancient Siberians, together with additional specific ancestries.

Another important consequence of the migration out of Africa was the exposure of modern humans to unfamiliar environments, including new and diverse climates ranging from frosty Siberia to high Himalayan mountains, unfamiliar foods, and pathogens causing new diseases. All these conditions represented challenges to survival (selection pressures). Those with the genetic makeup which would enable them to cope best with the local conditions and produce offspring who were also well equipped to survive would have been favoured. Adaptation could involve technological innovations such as developing suitable clothes, but sometimes genetic differences between humans could make the difference between survival and death—this is natural selection. As different traits were favoured in different environments, people living in different parts of the world have developed different local adaptations, some of which are reflected in our appearances today (see Bigger picture 6.2, Fig A). To find out more about how selection has shaped our health and looks, see the section 'Natural selection and sexual selection' below.

Bigger picture 6.2
Race, racism, and genetics

Humans love to classify things, so it is no surprise that we classify ourselves. Scientific classification began with Linnaeus, a Swedish botanist who in 1758 devised the system we still use today to name every living thing (see Fig B), calling us 'Homo sapiens' (wise humans). He also subdivided humans into several types. Some we might expect, such as europaeus

Fig. A Humans from different parts of the world. We may look different, but genetically we are all very much the same species.

6.2 a (i) meunierd/Shutterstock.com; 6.2 a (ii) Cultura RM/Alamy Stock Photo; 6.2 a (iii) Cultura RM/Alamy Stock Photo; 6.2 a (iv) Cultura RM/Alamy Stock Photo; 6.2 a (v) Jon Arnold Images Ltd/Alamy Stock Photo.

(Europeans) or *afer* (Africans). But others, such as *monstrosus* who, according to Linnaeus, included 'large lazy Patagonians' are just bizarre—what did he have against Patagonians who are neither particularly large nor in any way lazy? Judgemental and unscientific classifications such as this were soon forgotten.

The division of humans into types ('races'), however, flourished. It had its complications: no one could agree on how many types there were, or who belonged to each. Opinions varied from two to 200, according to one count. There were some quaint distinctions: 'the average standard of the Lowland Scotch and the English North-country men is decidedly a fraction of a grade superior to that of the ordinary English' wrote Francis Galton (Charles Darwin's cousin) in 1869. Two butterfly-fanciers once classified their favourite butterfly, the Chalkhill Blue (see Fig C), into well over 200 types like *fowleri, semifowleri, ultrafowleri, fowleri-margino, semifowleri-margino* and so on: a classification discounted by everyone else. Why don't we just ignore nineteenth-century human 'races' in the same way?

In humans, 'race' became entangled with colonialism, slavery, and other forms of exploitation. According to Wikipedia, racism is 'the belief in the superiority of one race over another, which often results in discrimination and prejudice towards people based on their race or ethnicity'. It is a social belief, near-universally condemned. But does it have a biological/genetic basis?

Races are discrete types. It is theoretically possible that any species (including humans) could be divided into races. With DNA sequences, we can test this possibility by comparing the sequences of people from different places. If Europeans and Africans belonged to different races, we would find genetic variants present at 100 per cent frequency in Europeans and 0 per cent in Africans, and vice versa. In fact, after examining the entire genome, we find zero variants of this kind, for these or any other supposed 'races'. There are genetic differences, of course, but they are quantitative in nature (eg 35 per cent frequency of a particular variant in one population, 46 per cent in

Fig. B Linnaeus worked out the basis of our modern classification system—but he didn't do so well when it came to humans!

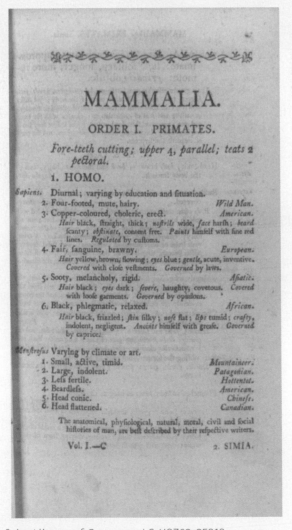

Courtesy of the Library of Congress, LC-USZ62-95219

Fig. C Chalkhill blue butterfly—once divided into over 200 'types'!

© Anthony Short

another). This gradation explains the confusion about the number of 'races', and so we write human 'races' in inverted commas. From a genetic point of view, there are no human 'races', and no scientific basis for racism. You may experience or observe the everyday reality of racism, but it is based on a purely social belief, not on a genetic reality.

❓ Pause for thought

If there are no human races, how is it that we can (often) tell at a glance where someone is from, or at least where their ancestors were from? How reliably can we do this? Consider ancestry-testing DNA companies—would you take a test? Why—or why not? What are the benefits or risks to you or your relatives?

From hunters and gatherers to farmers

So far, we have learned a lot about biological evolution, but human behaviour, lifestyle, and culture have also changed throughout our evolutionary history. Sociocultural practices have not only evolved alongside the genome, but also impacted genetic evolution itself.

Early human societies all practised hunting and gathering, which means they lived in small, mobile groups and relied on animal and plant food they could scavenge, catch, or find. This was a highly successful way of life, well suited to an unpredictable and ever-changing environment. There are still human communities following such lifestyles: indigenous groups inhabiting remote parts of Africa, Asia, Australia, and the Americas.

However, around 10,000 years ago, the climate changed, becoming both warmer and more stable. In these circumstances, a different way of life became possible: growing plants and looking after animals. This change is often called the **Neolithic** transition (literally meaning 'new stone' because of a change in stone tool technology in some but not all Neolithic societies), but more importantly involving the transition to food production. People learned how to cultivate plants, grow crops, and domesticate animals, which marked the beginning of agriculture and pastoralism fuelled by the favourable climate. Such a dramatic transformation of lifestyle from hunting and gathering to farming happened multiple times independently in different parts of the world: the Fertile Crescent in West Asia, China, Africa, New Guinea, and the Americas at different times (see Fig 6.10). It had profound consequences for both the animals and plants, and for us.

Farming expanded from these centres to envelop most of the world. Did this occur by movement of the farmers, who replaced the hunter-gatherers, or by hunter-gatherers themselves adopting farming? This question has been studied in detail in Europe, using archaeology to identify the spread of farming from the Fertile Crescent, and aDNA to study the genetic origins of the farmers. Farming came to south-east Europe from present-day Turkey around 8,000–9,000 years ago, but took several thousand years to spread throughout Europe, reaching northern Europe only around 4,000 years ago. Ancient DNA has shown that the early southern European farmers genetically resembled Near Easterners, but the first northern European farmers 4,000 years later were mixtures of Near Eastern farmer and European hunter-gatherer ancestries. Thus, present-day Europeans carry both of these ancestries (as well as large genetic components from later prehistoric migrations).

Fig. 6.10 Origins of agriculture—agriculture originated independently in many different parts of the world at different times in prehistory

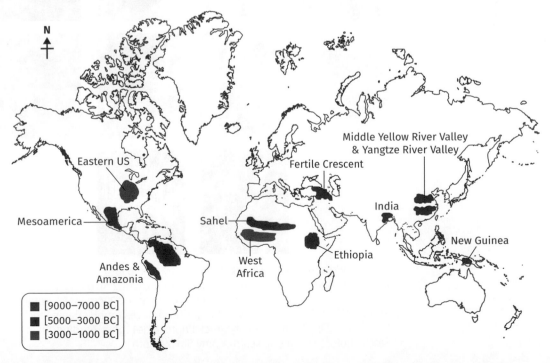

Choosing animals and plants with the most desirable traits over many generations resulted in modified forms: breeds and varieties more accustomed to humans, less toxic or with higher yields, such as domesticated animals and plants (see Fig 6.11). Food production increased the abundance of resources, so supported bigger communities and led to a settled way of life, with many people needing to live throughout the year alongside their domesticated animals and plants. People started to make radical changes to their surrounding environment, by cutting down forests to make space for fields to grow crops, or intensive irrigation projects to provide water. Human population densities became higher, and specialization and division of roles in the society were possible.

There were some down sides too, as the new lifestyle imposed new challenges. Disease epidemics spread rapidly in large settlements, and new diseases were transmitted from domesticated animals to the people taking care of them, for example influenza, smallpox, and measles. Agriculture also revolutionized the human diet, with new food sources becoming available. However, some argue that it narrowed the diversity of the food consumed and that a cereal- or rice-dominated diet worsened human health because of nutritional deficiencies, or conditions such as obesity and diabetes.

Fig. 6.11 In each of these pairs of wild vs domesticated animals and plants, the wild species are numbered 1 and the domesticated numbered 2. You can see the difference selective breeding has made!

6.11a1 Helen Kattai/Alamy Stock Photo; 6.11a2 Emilio Ereza/Alamy Stock Photo; 6.11b1 John Cancalosi/Alamy Stock Photo; 6.11b2 Arco Images GmbH/Alamy Stock Photo; 6.11c1 imageBROKER/Alamy Stock Photo; 6.11c2 Volodymyr Burdiak/Alamy Stock Photo; 6.11d1 mauritius images GmbH/Alamy Stock Photo; 6.11d2 blickwinkel/Alamy Stock Photo; 6.11e1 robertharding/Alamy Stock Photo; 6.11e2 mediasculp/Alamy Stock Photo

Such profound changes in the surrounding environment and lifestyle sometimes resulted in adaptive genetic responses by people to the new foods and new diseases. The best-characterized human adaptation acquired during these times is **lactose** tolerance. Humans, like other mammals, traditionally drank milk only in infancy, and the ability to digest lactose (the sugar in fresh milk) declined with age, so that adults would experience unpleasant side effects if they drank fresh milk. But as milk from domesticated animals became an additional and renewable food source in pastoralist communities, adults who could consume milk were at an advantage, and a genetic variant enabling its digestion in adulthood without unpleasant consequences became common in those populations (see the section 'Natural selection and sexual selection' below).

Environmental transformation, attachment to land, surplus food and its storage, the accumulation of goods, tools, and knowledge, and further technical innovations provided the basis for changes in social behaviour, leading to ownership of land, and complex social organization including **hierarchies** with centralized administration. All of these changes laid the foundations for a remarkable human achievement that has subsequently affected everyone for good or ill—civilization.

Natural selection and sexual selection

Different genetic variants can have different consequences. Most have no effect at all (and so are called neutral), while some might be detrimental and decrease the chance of survival and reproduction. Such deleterious variants are thus removed from the population over the generations, a process called **negative** or **purifying natural selection**. Occasionally, variants may be beneficial, for example by making the carrier more sexually attractive or better adapted to the environment. These variants increase the chance of having offspring and thus increase in frequency in the population—so-called **sexual selection** or **positive natural selection**, respectively. A very few variants are advantageous when present in a single copy in an individual but cause disease when present in two copies—the phenomenon of **balancing selection**, illustrated by the **sickle cell allele**. When present in two copies it produces red blood cells of unusual shape that get stuck in blood vessels and cause severe anaemia. Carriers having only one copy, on the other hand, do not develop the disease, but are less likely to suffer from malaria. The red blood cells of carriers are more toxic to the malaria parasites, and if they get infected, are more easily eliminated. This results in fewer parasites and thus less severe symptoms.

Positively selected variants are particularly interesting since they are responsible for the evolution of novelty. This is the main type of selection that has made us different from other great apes, and driven human differentiation around the world. These variants can be identified by comparing the DNA of humans and their close relatives (great apes and Neanderthals) or by comparing the DNA of diverse populations around the world with one another. They stand out even if we do not know in advance anything about their particular selective advantage.

Different sets of traits are favoured in different environments, just as different clothes are better suited to different weathers. Searching for genetic variants common in one geographical region, but rare elsewhere, led to the identification of variants in the genes *SLC24A5*, *SLC45A2*, *TYR*, *BNC2*, and *OCA2* contributing to light-coloured skin in Europe and Asia. It may be that the lower levels of sunlight and UV exposure led to lower vitamin D production in dark skin, consequently raising the risk of the harmful condition rickets (natural selection), or alternatively that light skin was attractive to partners (sexual selection). Interestingly, skin lightening happened relatively recently, within the past 10,000 years: earlier Europeans, including Britons, had dark skin and hair. By that time, blue eyes were already widespread in Europe, as seen in the Cheddar Man. There is no obvious way in which blue eyes might have helped survival, but perhaps such novelty was seen as attractive and improved competition for mates.

Other known adaptive traits include increased height in northern Europe and the lactose tolerance in adults mentioned earlier in the section. Various other new characteristics including the response to alcohol, thick, straight hair, and dry ear wax in Asia (see Fig 6.12) show the genetic signatures of recent positive selection, but the reasons for selection range from speculative (thick, straight hair may have been attractive to partners) to completely unknown (dry ear wax, Fig 6.12). Finally, mixing with Neanderthals and Denisovans has also provided humans with adaptive variation. The most striking example of such adaptive introgression is a Denisovan version of the *EPAS1* gene found today in people who live at high altitude in the Himalayan mountains, such as Sherpa and Tibetans. This version is expressed at lower levels, in turn improving oxygen transport and decreasing the risk of altitude sickness, making it easier to live in this harsh environment.

Are we still evolving?

One of the questions we, as scientists working on human evolution, are most frequently asked is: 'Are we still evolving?' The thinking behind the question is often that evolution depends on 'the survival of the fittest', so in a world where survival and reproduction are supported by favourable environments, abundant food and ever-improving medical technologies, survival rates are high and there is less opportunity for natural selection to act. You will appreciate straight away that this thinking is irrelevant to

Fig. 6.12 People in most parts of the world have wet ear wax (a), but in East Asia there has been strong selection for dry ear wax (b)—but we have no good idea why!

(a)   (b)

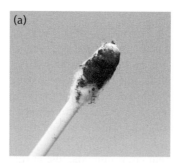

(a) kzww/Shutterstock.com; (b) Shaynepplstockphoto/Shutterstock.com

much of the world, where starvation or infectious diseases such as malaria and HIV remain tragically high. In those circumstances, evolution continues in the most brutal way. But is it applicable to even the most pampered sections of developed countries? In such societies, not everyone contributes equally to the next generation. Mate choices and decisions about the number of children remain major factors. Research on online dating patterns has revealed that people tend to seek partners with similar ethnicity or educational achievement, while they may favour the best possible physical attractiveness or income. So, evolution continues here as well. And the biggest uncertainty is the prospect of modifying our own genomes in ways that would be transmitted to our children. This is ethically unacceptable now, but will this still be true in 100 or 1,000 years' time? Watch this space!

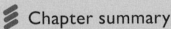 Chapter summary

- Information about human evolution has traditionally come from studies of living humans and apes, their fossils, and archaeology, but is now coming increasingly from genomics, including genomic studies of aDNA.
- Humans are one of the great apes, most closely related to chimpanzees and bonobos, then to gorillas and then orang-utans; our common ancestor with chimpanzees and bonobos lived 5–7 million years ago.
- The human genus *Homo* evolved in Africa more than two million years ago and our species *sapiens* also originated in Africa, around 300,000 years ago.
- People outside Africa originated from the migration out of a small African population at around 55,000 years ago, followed by expansion into most of Asia, Australia, and Europe by 45,000 years ago and into the Americas by 15,000 years ago.
- During these expansions, they mixed with other forms of early human, Neanderthals and Denisovans, so that all non-Africans carry about 2 per cent of Neanderthal DNA, which influences our characteristics such as hair and skin colour, virus resistance, and diabetes risk.
- After 10,000 years ago, farming arose independently in several different parts of the world, leading to domestication of animals and plants, and increases in population size. It later provided the potential for all the changes we call 'civilization'.
- People always adapted to their local conditions, leading to the diversity of humans we see today around the world. The differences from place to place are, however, gradual and there are no genetically distinct races of humans.
- Striking examples of local genetic adaptation whose genetic basis is understood are the lighter skin colour seen in Asia and Europe, and the ability of adults in some places, but not others, to digest fresh milk.

- People sometimes suggest that human evolution has slowed down or come to an end, but evolution continues because new pathogens are appearing continuously, and choices about partners and numbers of children mean that each generation is slightly different.

Further reading

1. The question of whether humans are still evolving: https://www.yourgenome.org/stories/are-humans-still-evolving.
2. The evolution of the human brain: https://www.yourgenome.org/stories/evolution-of-the-human-brain.
3. Jared Diamond, *The Rise and Fall of the Third Chimpanzee*, Harper Collins (1991). We humans can be considered a third chimpanzee, alongside the common chimpanzee and bonobo.
4. Yuval Noah Harari, *Sapiens: A Brief History of Humankind*, Harper (2011). The title says 'brief', but the book is 456 pages long: a well-deserved bestseller.
5. Svante Pääbo, *Neanderthal Man*, Basic Books (2014). The autobiography of the leader of the ancient DNA field, including a good explanation of the science and technology.
6. Guido Barbujani, *Human Races*, https://www.cell.com/current-biology/fulltext/S0960-9822(13)00027-4. Questions about race that might occur to you, and the answers of a geneticist.

Discussion questions

6.1 A critically endangered population of gorillas is barely surviving in a small remnant of tropical forest. The people living in the surrounding region desperately need to cultivate the forest to produce food, which will drive the gorillas to extinction. You have to recommend what should happen. What would you suggest as the best way forward?

6.2 Imagine that in the future, artificial intelligence has developed so far that it could reliably predict the partner who would provide the best relationship for you, as long as you supply all the details of your background, likes and dislikes, genome sequence, and so on. Would you share this information and follow the recommendation? Try and explain your responses.

6.3 Imagine that the genome of an unborn baby can be determined at a very early stage and any DNA position can be edited with a 99 per cent chance of introducing the intended change, but a 1 per cent chance of creating a random change instead, which will sometimes be harmful. A baby is found to carry a genetic variant that has severely harmful consequences. Would you go ahead with the editing? What information would you consider in order to make this decision? Would your decision be different if the genetic variant was only mildly harmful, or the desire was to 'improve' a normal baby? Discuss the implications of these types of choice for people in the future.

GENOMICS: READING AND WRITING GENOMES

7

In the first six chapters of this book you have explored some of the amazing advances that have resulted from our growing understanding of genomics. Who would have imagined, only a few years ago, that we would be able to identify the genetic causes of rare diseases affecting only a few people in the entire world? That we would be able to track the progress of a cancer evolving drug resistance and change our treatment accordingly, or that we would be able to identify the origins of an outbreak of infectious disease to a single individual? No one would have believed scientists if they had said they would be able to identify Neanderthal DNA in modern humans, or that it would be possible to demonstrate evolutionary links between species which look entirely different from each other. The idea of driving genetic infertility through a population of disease-causing insects, or of altering the genetic makeup of individuals to treat previously incurable diseases, would have seemed to be the stuff of science fiction, not science fact.

Yet all of these things are now possible as a result of our ever-growing understanding of both the human genome and the genomes of thousands of other species which range from the compact genomes of microorganisms to much bigger genomes of huge animals like blue whales and many of the plants we rely on for our food.

All of this knowledge depends on the incredible advances which have taken place—and are still taking place—in the technologies involved in DNA sequencing and in DNA manipulation. As you will see in this chapter, the technology needed to read a genome has changed from huge machines that needed many scientists to work alongside them, taking years to read a single genome, to small units which can be left alone to read a genome almost overnight. Our ability to handle the enormous amounts of data generated has developed alongside the sequencing technology that creates it. Scientists are also working on ever more sophisticated tools to enable us to manipulate genomes and to produce longer and longer pieces of synthetic DNA. The development of the technologies that are powering the genomic revolution is one of the fastest moving and most exciting areas of biology today (see Fig 7.1).

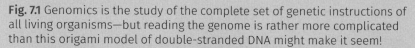

Fig. 7.1 Genomics is the study of the complete set of genetic instructions of all living organisms—but reading the genome is rather more complicated than this origami model of double-stranded DNA might make it seem!

yourgenome, Genome Research Limited. CC BY 4.0

Reading DNA: a definition of sequencing

DNA sequencing is the process of reading or working out the exact order of bases in a strand of DNA. DNA is a very complex molecule (see Fig 7.2) and the three-dimensional structure depends on the presence of four bases (adenine, cytosine, guanidine, and thymine), which are usually written in text as A, C, G, and T.

In most cells, long DNA sequences form molecules arranged in structures called chromosomes. These vary in size from about 50,000,000 to 300,000,000 base-pairs in the human genome. Each human has forty-six chromosomes (twenty-three pairs), which is about 3.2 billion bases in total. Current technologies don't allow whole chromosomes to be sequenced at once, so the DNA has to be broken down into smaller pieces, usually described as fragments. The number and order of the bases is then identified in each of these fragments, which can then be compared or overlapped using computers. The earliest DNA sequencing methods were expensive, slow, and complex. However, newer methods are much cheaper, faster, and easier to use.

Another way of understanding DNA sequencing is that the process can be used to convert properties of 'real' biological molecules into 'imaginary' molecules that can be stored, analysed, and manipulated using computers. Researchers can then make virtual edits to the DNA sequences, so that they can introduce the modifications into living cells to help understand what these sequences do and when they are used by the cells.

Fig. 7.2 DNA is a molecule which is both elegantly simple and yet highly complex—we are still in the early days of understanding all the secrets it holds.

sciencephotos/Alamy Stock Photo

Sanger sequencing: the first method for reading DNA

The Sanger sequencing method was developed by Fred Sanger in the 1970s. This method was the first that allowed scientists to read the DNA, and it was based on the processes used in cells to replicate DNA. What are the key steps in DNA replication?

First, the double helix of DNA is separated into single strands by enzymes, a process known as denaturation. You can also imagine this as the DNA being 'unzipped'. Once unzipped, the two strands can act as templates for creating two new strands of DNA. A short piece of RNA, called a primer, binds to the template strand, and is then bound by an enzyme called DNA polymerase. The DNA polymerase starts to make a new strand of DNA by incorporating free nucleotide bases (A, C, G, T) that base-pair with those on the existing strand. This process continues until two complete strands are made. An important detail is that the bases in the primer RNA molecule have to be able to bind or 'base-pair' with the bases in the template strand, following the same base-pairing rules that allow two DNA strands to form a double helix (see Fig 7.3).

Fig. 7.3 A summary of the process of DNA replication in the nucleus of the cell.

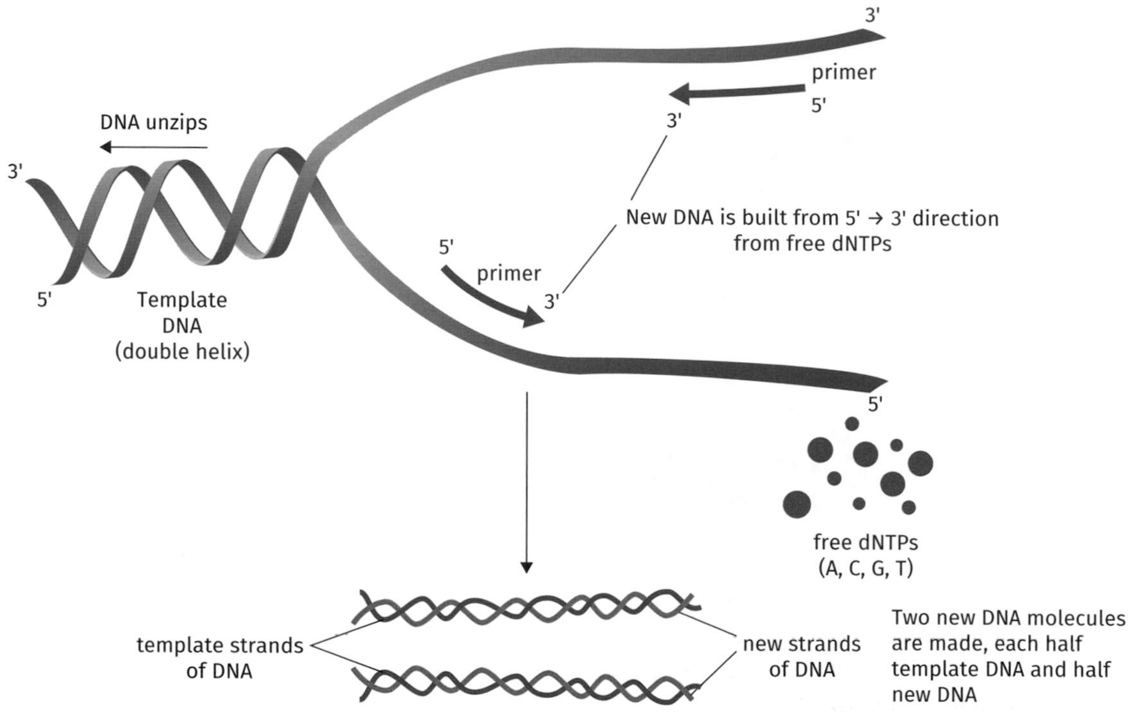

How is this process used for Sanger sequencing? As with DNA replication, the section of DNA that is going to be sequenced needs to be unzipped to be able to act as a template. In the lab, this is done by heat or adding chemicals. In the reaction mix there are primers (this time made from DNA), DNA polymerase enzymes, and nucleotide bases. At least one of these bases is radioactively labelled so that the DNA that is being newly synthesized can be detected at the end of the process.

Terminator molecules are crucial in Sanger sequencing: the incorporation of a terminator into a newly made DNA stops any further bases being added to the DNA strand. We add small amounts of alternative versions of the nucleotide bases, known as 'terminators', into the reaction mix, which can't be extended any further because they lack the hydroxyl ('OH') group that is required to chemically attach the next nucleotide (see Fig 7.4). There are specific versions of terminators for each of the four bases, so when an A terminator is used, the last base of the DNA strand will always be an A. If a C terminator is used, the last base in the DNA strand will always be a C. The same principle is also true for G and T terminators. The process is repeated multiple times so, if the concentration of terminator is low enough, and there is enough template DNA in the reaction, there should be DNA strands that represent every possible stop point (so every base) within the whole DNA sequence.

Fig. 7.4 Compared to dNTPs, terminators (ddNTPs) lack the hydroxyl (OH) group that is necessary for the chemical reaction that could add another base to the DNA molecule, meaning that addition of a terminator molecule stops the DNA strand being extended further.

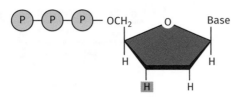

Dideoxynucleotide (ddNTP)

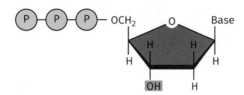

Deoxynucleotide (dNTP)

In the Sanger sequencing process, versions of this terminator reaction are set up separately for each of the four bases, as the same radioactive label is used to 'tag' each base. The result of each of the four reactions is a set of pieces of DNA of different sizes, but each with each fragment ending with the same base letter as the terminator included in the reaction. These four reactions are then loaded onto separate lanes of an acrylamide gel, where DNA pieces can be separated by size using a process called electrophoresis. If the fragments can be placed in size order, and compared between the four reactions, it is then possible to work out the complete sequence of the DNA molecules that were the templates of the reaction. Fig 7.5 summarizes the process.

How does electrophoresis allow the fragments to be ordered by size? The acrylamide gel is effectively a block of firm, thin, clear, chemically inert and electrically neutral jelly. Once the four reactions are each loaded into adjacent holes in the gel, an electrical current is applied. DNA molecules are negatively charged, so are drawn towards the positive electrode, and move through the gel towards it. The smaller fragments 'wriggle' through the gel more easily than the large fragments, so the fragments begin to separate by size. The gel is run until the range of fragment sizes expected is spread out across the gel—the results are much less useful if the smaller ones have fallen off the end of the gel or if the larger ones are all clumped tightly together.

The results of the gel and the reactions can be seen by exposing an X-ray film to the gel. The radioactively labelled DNA bases emit X-rays that can make an image, called an autoradiogram. The darkness of the image

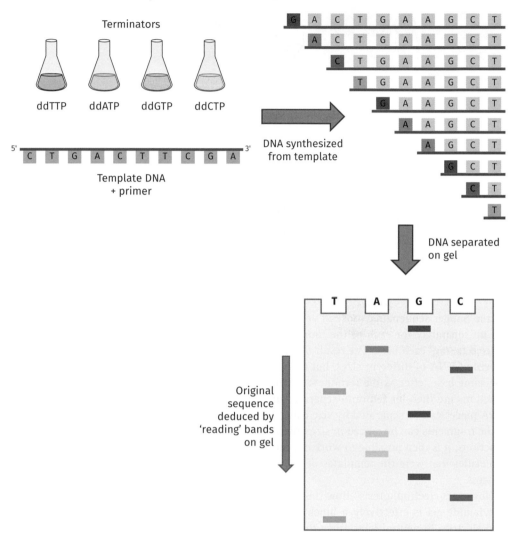

Fig. 7.5 The Sanger method of DNA sequencing. Although this has now been superseded by much faster, cheaper ways of analysing DNA, almost all modern methods have their roots in the original Sanger method. Colours are used to make the diagram clearer, as all the radio-active markers are actually detected as dark bands.

represents how much DNA is there. Where DNA molecules of the same size travel together through the gel, a dark band appears on the autoradiogram. Each band of the gel corresponds to size of molecule where one of the terminators has been added, so where a specific base was present in the template sequence. If all four reactions are run side by side it is possible to read off the DNA sequence, as there should be a band representing every base in the template.

We explore one of the biggest milestones in the use of DNA sequencing—the sequencing of the human genome—in Case study 7.1.

Case study 7.1
The Human Genome Project

The Human Genome Project was a huge effort by scientists across the world to read the entire human DNA sequence and to make the whole DNA sequence publicly available to all researchers without restrictions. The project took about fifteen years, and it spurred on the development of DNA sequencing technologies. At the same time, there were parallel efforts by a private US-based company. They wanted to make the human genome sequence something owned commercially, which would have restricted access to researchers. Fortunately for us all, publicly funded teams from countries including the US, the UK, France, Germany, Japan, and China, working in parallel and sharing data, managed to unravel the human genome first!

The UK home of the Human Genome Project was the Wellcome Sanger Institute (known then as the Sanger Centre), found in the countryside near the city of Cambridge. This institute, led by John Sulston, sequenced around 30 per cent of the whole human genome—that's around 850 million base pairs! Working together, the different teams completed the Human Genome Project in 2001, with a gold standard genome published in 2003. Prior to the completion of the genome sequence, it was not possible to be sure how many protein-coding genes were in the human genome; around 22,000 were identified when the genome was released. This is similar to other mammals—and perhaps fewer than some might have expected!

The significance of this scientific endeavour was highlighted by a trans-Atlantic press conference led by US President Bill Clinton and UK Prime Minister Tony Blair, who described the project as:

> a breakthrough that opens the way for massive advancement in the treatment of cancer and hereditary diseases. And that is only the beginning.

That final point is quite critical—it was only the beginning. John Sulston wrote in his book, *The Common Thread*, on the completion of the Human Genome Project:

> We are right at the beginning, not the end.

Completing the human genome is one thing; translating that into actual health-care advances and benefits is another, and something that takes a very long time. This work continues at the Wellcome Sanger Institute—which has grown into the enormous complex you can see in Fig A. It is now home to hundreds of scientists working on many different aspects of genomics—including all of the authors of this book. Many of the research projects undertaken at the institute relate to using a knowledge of DNA sequences to better understand questions that have a direct impact on human health. And now thousands of patient genomes are sequenced each year in the DNA sequencing pipelines, using high-throughput, short-read-based technologies—something which would have been hard to imagine when the institute was founded!

Fig. A The Wellcome Genome Campus.

Capillary sequencing: speeding up sequencing to factory scale

The Human Genome Project led to a need to scale up DNA sequencing technologies, leading to the widespread use of capillary sequencing (see Fig 7.6). This process is based on Sanger sequencing, but with a number of modifications to speed up the process and make it cheaper. The improvements in processing time that these changes in technology enabled allowed far more samples to be processed by the same number of people, and powered the progress of many large-scale genome projects, including the study of organisms frequently studied in laboratories, such as mice, flies, and worms, and pathogen genomes, such as *Plasmodium falciparum*, the parasite that causes most human malaria.

Rather than using radioactively labelled bases, the terminators used in capillary sequencing have coloured tags that can be made to fluoresce (glow). These terminators can be incorporated into a DNA strand and be detected by a camera. It is possible to make four differently coloured tags, which means that all four bases can be included in the same reaction mix, rather than in four separate reactions. This both decreases cost and increases the number of samples that it is possible to run.

Running the gels for size separation was also made faster by new technology. Instead of having to cast lots of solid gels, liquid gels arranged side by side in long, thin 'capillary' fibres were used to separate DNA, in a process controlled by a machine. These small fibres allowed the use of very high electric fields, which means the DNA fragments could be separated much faster than with solid gels.

How does the capillary sequencing process work? A sample of DNA is injected into a capillary, which is done by dipping the capillary and an electrode into a solution of the sample, and then briefly applying an electric

Fig. 7.6 A capillary sequencing machine (a) results in a trace like the one shown in (b), which can be used to give the order of bases in a DNA sequence.

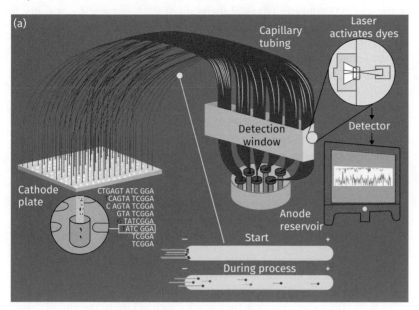

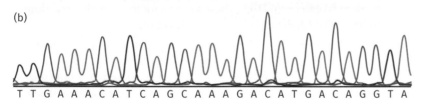

T T G A A A C A T C A G C A A A G A C A T G A C A G G T A

(a) DOE Joint Genome Institute, Lawrence Berkeley National Laboratory, University of California

current. As the DNA is negatively charged, this current causes the DNA to move into the capillary. The smallest fragments move fastest through the gel, and the fluorescence from the terminators is the first to be detected by the camera. The setup is calibrated so that fluorescence from each of the bases in the DNA fragments can be detected in turn, by making sure that the gel moves past the camera at the correct speed. The fragments that are the same size move through the gel almost simultaneously, so the fluorescence is detected almost at the same time, forming bursts of detected fluorescence.

If these bursts are plotted on a chart, they appear as peaks. The level of each colour shows which base was detected (plotted on the vertical axis), and the order in time in which the bases are detected (plotted on the horizontal axis) matches the order of bases in the sequence. The maximum length of a continuous sequence obtained using this method is around 800 bases. This text sequence of DNA bases is known as a 'read' (pronounced to rhyme with speed).

High-throughput sequencing with short reads

The further development and application of new sequencing technologies in the late 2000s hugely changed the scale of DNA sequencing projects. It became possible to generate millions of sequencing reads at the same time for pennies and fraction of pennies, which is a huge contrast to the scale of tens or hundreds of capillary reads costing multiple pounds each.

Multiple of these 'high-throughput' sequencing technologies were developed and sold by different companies, including Illumina (sequencing by synthesis technology), ABI (SOLiD technology), Roche (454 technology) and Helicos (direct sequencing). The most widely used of these technologies by far in 2019 was the one sold by Illumina, for reasons including cost, quality of reads, and the practical fact that many researchers adopted and used the technology that was accessible at their institution. Market forces are very important in genomics, particularly for large-scale projects. It would be a poor investment for a research institute, university, or company to buy a machine which is expensive to run or that might become obsolete in months rather than years.

Some features are common to all of these high-throughput technologies:

- Upfront fragmentation of the long molecules in the original DNA sample. These molecules are shattered into many thousands of pieces per genome, usually a few hundred base-pairs long. Specific synthetic DNA sequences are then stuck ('ligated') on to the end of these random fragments of DNA. These act as handles to enable reactions to be carried out on all of the molecules at once.

- The reads generated by these technologies were initially very short—a few tens of bases, though current versions can be hundreds of bases. Given the millions of pieces of data generated, it is effectively impossible to analyse these reads manually. Instead, computers and algorithms are used either to match these reads to a known sequence (a process known as mapping) or to overlap the short reads to make a much longer sequence (a process known as assembly). It is also possible to pick out specific genes of interest (such as those known to be involved in a particular process, like insulin production or onset of cancer) by using synthetic DNA sequences that will only base-pair with the genes that have a complementary sequence.

What happens after the synthetic sequences are added to the ends of the DNA molecules? This varies between the technologies, so we will focus on the widely used 'sequencing by synthesis technology' sold by Illumina in Scientific approach 7.1.

Long-read sequencing

While some biotechnology companies have focused on generating lots of cheap and short sequencing reads, other companies have taken a different approach, and have developed technologies that allow very long DNA molecules to be captured in single reads. The definition of 'very long' varies, but the current long-read technologies allow molecules that are tens of thousands of bases long to be read routinely. There are reports of reads captured which are more than two million bases long, which is not that much smaller than most bacterial genomes.

Scientific approach 7.1
Illumina sequencing by synthesis technology

In the Illumina technology, the synthetic DNA sequences added on to the ends of DNA fragments are known as adaptors. Before the DNA is loaded onto a sequencing machine, the DNA is denatured to make single-stranded molecules with adaptors at both ends. The DNA bases within these adaptors can then be made to bind to complementary DNA sequences, which can be used to capture the single-stranded molecules to the complementary sequences. Many copies of these complementary sequences are stuck to the glass surface in an enclosed channel within a glass microscope slide—a device which is known as a **flow cell**. Why is it useful sticking these molecules to this glass surface? It enables the binding of modified, fluorescently labelled terminators to be captured with a camera, and the sequence to be captured within sequencing reads. Unlike capillary sequencing, all of the nucleotides used in Illumina sequencing are fluorescently labelled.

After the single-stranded DNA molecules are captured on the flow cell, a series of copies of the molecule are made that bind right next to the original template molecule, which are called '**clusters**'. This is analogous to the leaves of spring-flowering bulbs being in a cluster around where the original bulb was planted. The process is similar to the polymerase chain reaction (PCR), but with the molecules stuck to the surface of the flow cell rather than in solution. The concentration of template DNA is carefully controlled to make sure that the clusters don't form too close together and overlap. This would be like planting many spring bulbs close together, then attempting to tell from a distance where one ends and the next one starts! Each one of these clusters is from a different template molecule, so it is equivalent to the samples read by capillary machines. But a crucial difference is scale—millions of these template molecules can be processed in one flow cell, while a capillary machine usually processes ninety-six at a time.

Once the clusters have been formed, the molecules form lots of adjacent template molecules to which nucleotides can bind. The fluorescently labelled terminator bases can bind with free complementary bases in each cluster, and the ones next to a specific part of the adaptor sequence will be incorporated into the DNA. After this single cycle of DNA synthesis, all the nucleotides that are not incorporated into the clusters are washed away. Then images are taken which cover all of the channels in the flow surface, producing images where the four fluorescent colours of clusters correspond to the colours of the four DNA bases (see Fig A).

On its own this wouldn't be that useful, as the fact that these nucleotides are all terminators means the sequence would only be one base pair long. But a chemical modification on these nucleotides means that the fluorescent part of the DNA terminator can be removed after the image is taken, effectively making it into a normal DNA base. This means that another batch of fluorescently labelled terminators can be incorporated into the DNA clusters, and another round of images taken. The process of DNA synthesis, washing, and imaging can be repeated for many rounds or 'cycles'. The images can be

Fig. A Illumina sequencing technology is widely used in research to investigate DNA.

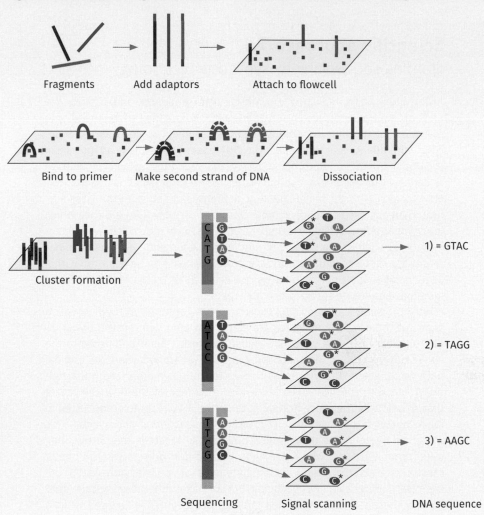

Adapted from Yuan Lu et al, 'Next generation sequencing in aquatic models.' *Next Generation Sequencing—Advances, Applications and Challenges*, 14 January 2016. DOI: 10.5772/61657. https://www.intechopen.com/books/next-generation-sequencing-advances-applications-and-challenges/next-generation-sequencing-in-aquatic-models. CC BY SA 3.0

stacked on top of each other using computers, with each stack of spots representing the sequence of the original template DNA molecule. Processing of these image files generates a text file that contains the sequences of each of the template molecules as individual reads. There are millions of reads from each flow cell, so the resulting text files are usually very large.

In the first version of the technology the sequencing reads were twenty-five bases long, but the current versions of this technology typically use reads that are 75, 100, or 150 bases long. These longer reads are easier to analyse than shorter reads, as there is more information to compare to a known genome sequence or to overlap with other reads.

Long-read technologies available now include those manufactured by Pacific Biosciences and Oxford Nanopore Technologies. They are based on different scientific principles, so have different benefits and limitations. The Pacific Biosciences technology uses imaging of fluorescent bases in a long DNA molecule to find the DNA sequence. A key difference compared to the Illumina technology is that the DNA is not copied first—fluorescence is detected from single, fluorescent DNA bases being made into a DNA strand. Not needing to copy the DNA first reduces errors, and makes it possible to sequence molecules that are too long to copy in a tube. However, it means that advanced imaging technologies are needed to capture the very small intensity of light generated from each base. This means that these sequencing machines are relatively large and expensive.

A different company makes sequencing machines that don't use imaging to read DNA. The Oxford Nanopore Technologies sequencing machines are small (smaller than a smart phone, though a little bigger than a USB drive), and contain membranes that are studded with protein channels, known as pores. These membranes separate two solutions of charged ions, which move through the pores. Long, single-stranded molecules of DNA can also move through these pores. As the DNA moves through the pores, it changes the ions that are able to flow through the channel at the same time. This produces a change in the electrical current at the pore, which can be detected by sensitive electronics near the pores. This technology produces the longest reads currently available, but has the disadvantage of more errors in the bases being detected than some of the alternatives. Some of these errors can be corrected by comparing multiple reads or combining with the reads produced by another technology. This technology is also the most portable of the DNA sequencing technologies currently available, with sequencing able to be done in remote places for disease epidemics or in remote environments to identify species within in an ecosystem (see Fig 7.7). This technology has also been used in space.

Fig. 7.7 Long-read sequencing is making it possible for scientists working in extreme environments, from rain forests to deserts, to identify species accurately as they find them.

© Anthony Short

Initially these technologies cost a lot more per base of DNA sequenced than the short-read technologies, but improvements in long-read technologies have reduced the relative costs and these long reads can have greater value than many more short reads.

Long reads can cover non-unique regions between unique regions, which are often impossible to put together using short-read technologies alone. These non-unique regions make up most of the human genome, and are found between genes. These long reads can also make it possible to tell apart genes which are very similar but have some unique sequences (many of these very similar reads are 'paralogues', which are multiple copies of certain genes that have been produced during evolution). These long reads can also make it possible to find unusual combinations of DNA that are very different from the combinations that are expected in most cells. A good example of this would be the DNA in cancer cells, where DNA is often swapped and duplicated between chromosomes. These different pieces of DNA are known as structural variants.

Short reads might miss these breaks, as only those reads exactly on the join would show the new sequence. The unusual combinations might not be in unique DNA sequences, making it harder to figure out where the break is in the genome. Longer reads can cover many more unique sequences either side of the new join, making it possible to identify more of this 'shuffling' of sequences on a large scale and so making it easier to pick out structural variants. Long-read sequencing has huge potential in understanding and managing both communicable and non-communicable diseases, in hospitals, laboratories, and out in the field (see Fig 7.8).

Sequencing to reveal active genes

When protein-coding genes are activated, an enzyme called RNA polymerase makes RNA molecules from nucleotides using a DNA template (a process called transcription) and these RNA molecules are used as guides to make protein molecules (a process called translation).

This means that detecting the presence or absence of specific RNA molecules within a cell can tell researchers which genes are active or inactive in a specific sample at a specific time. This can help researchers find the genes involved in a specific process, such as cell division or DNA replication. It can also help to identify which genes are important for different tissues, such as what makes brain cells different from liver cells. Detecting RNA molecules can also give clues about what is happening in a specific set of conditions—for example, which genes change activity when cancer cells are exposed to a drug. Detecting the sequences of all of the RNA molecules in the cells would help to form a complete picture of what genes are active in the cells.

Unfortunately, most DNA sequencing technologies can't directly read RNA. However, an enzyme is used to convert the sequence of the RNA molecule into a complementary single-stranded DNA (inventively called cDNA), which can be read by all DNA sequencing technologies. This use of an RNA template to create a strand of DNA is called 'reverse transcription',

Fig. 7.8 Portable equipment made by Oxford Nanopore (a) enables scientists to sequence the genome of the Ebola virus during an outbreak in Guinea (b), allowing scientists and doctors to track the sources of the disease and contain the outbreak.

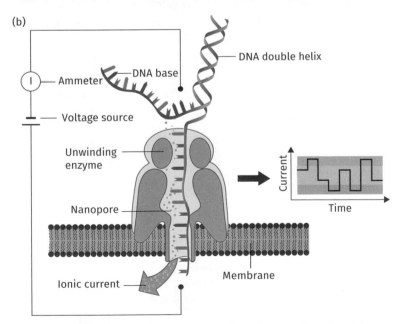

(a) https://www.iflscience.com/health-and-medicine/how-small-backpack-fast-genomic-sequencing-helping-combat-ebola/. (b) Adapted from Göpfrich, K and Judge, K, 'Decoding DNA with a pocket-sized sequencer.' *Science in School*, issue 43: Spring 2018, p 14. https://www.scienceinschool.org/content/decoding-dna-pocket-sized-sequencer CC BY

and uses a type of enzyme originally found in viruses that convert between RNA and DNA forms.

These cDNA molecules can then be sequenced and counted, to identify which genes are active in the cells. This sequencing approach is called RNA-seq, and can be used by both small and large projects.

The ENCODE project

One large-scale project that used this technique is the ENCODE (ENCyclopedia Of DNA Elements) project, which involved many human tissues and a large international team of scientists. Using the RNA-seq data from all these different cell types, they were able to reveal a lot about which parts of the genome were being used in different parts of the human body (see Fig 7.9).

Fig. 7.9 The principles of RNA sequencing.

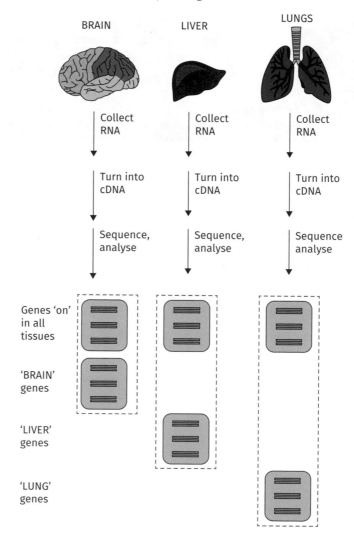

The aim of the ENCODE Consortium was to pick up where the human genome project left off, and to understand which parts of the genome that don't encode protein sequence actually have effects on the amounts of RNAs or proteins in human cells. The project ran from 2003 until 2012, with a large team of 1,640 researchers, using twenty-four standard types of experiment, which they used on 147 different human cell types. The analysis of this data revealed that 80.4 per cent of the human genome displays some kind of active function in at least one cell type—that's a lot more than the ~1 per cent that encodes proteins! The parts of the genome known to have these extra kinds of functions can be compared to databases of variable DNA in populations of people with and without disease to track down specific sequences or bases that might contribute to a risk of getting a particular disease.

Sequencing single cells

Technologies to capture and sequence DNA and RNA are now advanced enough that it is possible to sequence material that comes from individual cells. It is necessary to amplify the material before sequencing to have enough to load onto a sequencer, which can introduces errors and biases into the sequence data, but with careful analysis a lot can be known about what DNA or RNA was in each individual cell.

As many hundreds or thousands of cells are collected from each tissue or patient, it costs both more time and more money to do single-cell sequencing than it would to mash up the whole tissue into a single sample. Is it really worth the effort? Researchers working on this area of biology would argue strongly that it is!

By being able to look at each individual cell, it is possible to find rare cell types that wouldn't be detectable by looking at the whole sample together. In human samples, an example might be the very rare drug-resistant cells within a tumour that would be the ones that would come back after drug treatment, or a rare kind of immune cell whose role is to signal to other more common types of immune cell. It is now becoming possible to compare samples from people with and without disease, to figure out whether it is particular and important specific cells that have gone 'wrong', or whether the relative amounts of different cell types aren't in balance. This type of information could be an important clue to working out what processes are key in the development of diseases where we currently don't understand the causes. For example, an international group of leading scientists has agreed to collaborate to build a Human Cell Atlas, collecting many thousands of human tissues to help build detailed 'maps' of all of the cells in the human body (see 'The Human Cell Atlas' below).

Single-cell sequencing can also be applied to entirely non-medical problems, such as understanding all of the microbes present in an ecosystem. Many of these kinds of organisms would be impossible to grow or purify in a laboratory, but by using technology that allows each cell to be captured independently we should be able to begin to take 'snapshots' of many different types of ecosystems.

The Human Cell Atlas

The Human Cell Atlas is a large international project that began in 2016, which aims to use newly developed single-cell technologies to chart and map the properties of all human cells, across all tissues and organs. The project will make much more detailed maps than previous projects: as new online maps have continents, countries, or cities and houses, the Human Cell Atlas will contain information about organs, tissues, and individual cells. The project also aims to capture different types of human diversity, so will include samples from different countries, ages, diseases, and ethnic groups. The datasets generated by the project will be enormous compared to other biological data sets, so huge amounts of effort are being put into generating the computer infrastructure that will support browsable and searchable databases that are openly accessible to researchers all over the world. It is hoped that the data generated by this project will be almost immediately useful to researchers studying human diseases, as it should identify which cell types are missing or different from the rest of the population not affected by the specific diseases.

Editing DNA sequences in cells

Now we can read DNA from cells, can we learn how to make changes to it? Multiple technologies exist that allow scientists to edit DNA within living cells. One of the simplest and most widely used is the 'CRISPR–Cas9' system, which has been rapidly adopted around the world. This elegant technique has made it possible to edit the genome with more accuracy than ever before.

There are two key molecules that are needed to introduce changes (also known as mutations) into the DNA sequence of interest, known as the target. An enzyme, Cas9, acts as a pair of 'molecular scissors' that can cut the two strands of DNA at specific locations in the genome. This enables short pieces of DNA to be added or removed. There is also a short piece of RNA that is called a **guide** RNA (gRNA). A short part of this sequence has RNA bases that are able to base-pair with the target DNA sequencing within the genome, while a longer sequence of the same RNA molecules acts as a scaffold, which binds to the Cas9 enzyme. The gRNA is able to make sure that the Cas9 enzyme cuts in the right place in the genome. In theory, the guide RNA should only bind to locations in the genome where there is perfect base-pairing. Once the Cas9 enzyme has cut the DNA, the cell is able to detect that the DNA is damaged, and then tries to repair it. At this step, the mutation can be caused by 'incorrect' repairs, with new DNA sequences incorporated into the original genome (see Fig 7.10).

Where does this system come from? Some bacteria use very similar gene-editing systems to respond to invading pathogens such as viruses. The CRISPR system allows the bacteria to snip out parts of the virus DNA and keep it in their genome, to help defend against the virus next time it attacks. Scientists adapted this system to work in other genomes.

Why would you want to edit the DNA within a cell? This precise altering of the DNA sequence allows scientists to understand more about the

Fig. 7.10 CRISPR–Cas9 technology is revolutionizing the process of gene editing.

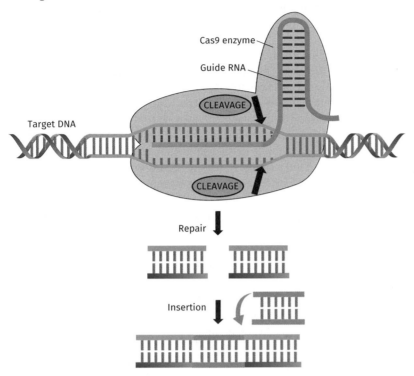

functions of genes and variants of a gene (almost like different spellings of a name, many genes have letters that vary between different people). Previous technologies did not allow such precise changes, and were slower and much more expensive. Some technologies introduced mutations randomly, which means that it was not possible to control where mutations occurred, and you would have to hope that you would make a mutation that was in the gene you wanted to study. Other more precise systems exist, such as transcription activator-like effector nucleases (TALENs) and zinc-finger nucleases (ZFNs), but the CRISPR–Cas9 system stands out as the fastest, cheapest, and most reliable.

What can we do with this system? There is a range of medical problems that have a genetic component—so, for example, the CRISPR–Cas9 system could be used to edit cells in the adult body to fight diseases such as cancer. This system has already been tried in a number of exceptional or life-threatening cases. More controversially, some people might suggest that such technologies should be used on germline (reproductive) cells, preventing genetic diseases in a way which would be passed on to future generations. Editing germline cells is currently illegal in most countries, including the UK.

It is important to remember that much of the research using this system is currently using animal models or isolated human cells (grown in flasks or dishes). A current focus is eliminating 'off-target' effects, where the Cas9 enzyme cuts at a different DNA sequence to the one intended. This can make results from isolated cells harder to interpret in a research project, but could have more serious consequences if it happened during treatment of a patient.

Most gRNAs consist of a specific sequence of twenty bases (or letters). These are complementary (perfect matches) to the target sequence of the gene that is edited. However, not all twenty bases need to be matches for the RNA to bind. If, for example, nineteen of the bases match somewhere else in the genome, edits may happen at the second unintended site. Better design of the gRNAs using knowledge of DNA sequences could help reduce the problem, in combination with using different variants of the Cas9–gRNA complex. Another approach uses a variant of Cas9 that will only cut a single strand of DNA instead of a double strand. This means that two Cas9 enzymes and two gRNAs need to act on the same DNA sequence for a cut to be made, reducing the chance of a cut being made in the wrong place.

Synthetic biology: writing entirely new genomes

As well as the small-scale edits possible with CRISPR–Cas9 technology, biologists have been developing tools to make writing DNA and genomes possible at a larger scale. At the moment, these technologies are limited to organisms with small and relatively well understood genomes, like bacteria and yeast. But in the future, it should be possible to use these kinds of DNA synthesis technology in more complex cells.

The technology that makes this 'synthetic' biology possible are tools to make and modify large pieces of artificial DNA sequences. For many years, researchers have been able to order and use short DNA sequences, like primers, that are tens of base pairs long. But newer technologies allow pieces of DNA that are hundreds of base pairs or longer to be made. Other technologies allow these pieces to be 'stitched' together to make simple chromosomes, and this has been done with bacteria and yeast.

But why would it be a good idea to make a 'synthetic' genome? One use is to see whether we fully understand what each bit of the genome does. If a specific sequence can be removed or modified, it means that it isn't essential for the cell in the conditions that you've tested. By doing this methodically within a gene or genome it is possible to check what many pieces of the sequences actually do. This makes it possible to begin to draw 'circuit' diagrams that capture how cells work, which look a lot like the diagrams found in computer programs or electronic circuits! Another application is to 're-program' the cell to do something new—an example would be to make a bacteria synthesize a drug that would be expensive or difficult to make using chemical processes alone. It would in theory be possible to re-program cells to make more dangerous drugs, but researchers in this area make a great effort to keep their work open to the public to help address these potential risks. Like many technologies, there are some potential risks that have to be balanced out against the great potential benefits.

Taking things forward

Looking to the future, it is an exciting time for genomic science. New discoveries are just around the corner that will really make an impact on health care in the twenty-first century. However, the field can't progress and evolve without fresh new talent. Genomics is a multidisciplinary field, involving lots of different skills sets, where some people specialize in lab techniques and others specialize in data analysis (which requires knowledge of coding languages), with some scientists dabbling in both. As genomic tests and innovations move into the clinic, more roles, such as genetic counsellors, will be needed to meet the demand, and more training in genomics will be needed for nurses and doctors. Hopefully by reading the chapters in this book you have seen the exciting range of opportunities that genomics offers—maybe you could be the next trailblazer in this field!

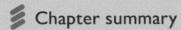 Chapter summary

- The invention of DNA sequencing technology enabled DNA to be read for the first time in the 1970s, using a process known as Sanger sequencing.
- The development of capillary sequencing helped to speed up many genome sequencing projects, as it is faster and cheaper than the previous methods.
- Further development of high-throughput short-read sequencing technologies has allowed many thousands of human genomes to be sequenced, along with other animal, plant, microbial, and viral genomes.
- Long-read sequencing enables complex sequences to be captured in one piece, enabling regions too complex for short reads to be analysed in detail.
- Genes that are active in particular cells or tissues can be identified by protocols that allow RNA to be used to make cDNA that can be read by DNA sequencers.
- Sequencing from single cells is now technically possible, and is allowing researchers to find the differences between cells from the same tissue.
- Editing DNA sequences in cells using protocols such as CRISPR–Cas9 allows researchers to change the DNA sequences within cells, so they can test what specific DNA sequences do within cells and potentially change faulty sequences or improve outcomes.
- Writing entirely new genomes is now possible for simple cells, as it is now possible to join together long synthetic DNA sequences.

Further reading

1. The Human Genome Timeline: https://www.yourgenome.org/facts/timeline-the-human-genome-project.

2. How the human genome sequence is being used: https://www.yourgenome.org/stories/how-is-the-completed-human-genome-sequence-being-used.

3. The Human Genome Project: https://www.genome.gov/12011238/an-overview-of-the-human-genome-project/.

4. The ENCODE project: https://ghr.nlm.nih.gov/primer/genomicresearch/encode.

5. The Human Cell Atlas website: https://www.humancellatlas.org/.

6. Genome editing: https://www.yourgenome.org/facts/what-is-genome-editing.

Discussion questions

7.1 What would be the challenges in using DNA sequencing technology to diagnose infections such as chest infections? How could you make sure samples didn't get mixed up?

7.1 Suggest some of the challenges involved in taking DNA sequencers to an extreme environment, like the jungle, desert, or the Arctic. How much equipment could you carry? How might you keep enzymes and other reagents cold?

7.1 Data sets for biology are getting bigger and bigger. What skills should we be teaching the next generation of biologists so that they can understand, analyse, and design these huge experiments?

GLOSSARY

A

Accessory genes Genes that are only present in specific strains of bacteria within a species.

Adaptive immune response A subset of immune response cells (B and T cells) that are produced to attack and destroy invading pathogen cells. The response relies on an immune memory based on previous exposures to the pathogen.

Adaptive introgression The incorporation of advantageous sequences from one species or population into the genome of another, for example from Neanderthals into the human genome.

Algorithm A series of steps or list of rules that enable a computer to perform a particular task in order to solve a problem.

Altitude sickness Illness which may result from exposure to low oxygen levels at high altitude, in people who are not born at altitude.

Amino acids The building blocks of proteins.

Ancient DNA (aDNA) DNA extracted from bones or teeth that are hundreds or thousands of years old. Usually present in tiny amounts (if at all), degraded, damaged, and contaminated with other DNA.

Angiogenesis The development of new blood vessels.

Animal modelling The use of animals such as mice, zebra fish, and flies to model human genetic diseases.

Anthelmintics Drugs that kill or paralyse parasitic worms and expel them from the host body.

Antibody A y-shaped protein produced by the immune system in response to invasion by pathogens.

Antigen Any substance that stimulates the immune system to produce antibodies.

Antigenic variation Process where parasites and other infectious agents change the proteins or carbohydrates on their cell surface to avoid a host immune response.

Apicomplexa A diverse group of mostly parasitic, single-celled eukaryotes, which all have an apicoplast (similar to a chloroplast but does not photosynthesize).

Array-cgh A new test for patients with learning and other disabilities which can detect small genetic changes.

Array techniques Molecular techniques used to detect chromosomal copy number changes on a genome-wide scale. Also known as microarrays and array-cgh.

Autism A condition affecting more than 1 per cent of the population, characterized by difficulties with social communication and social interaction, and restricted and repetitive patterns of behaviours.

B

Balancing selection A process which maintains advantageous diversity in a population.

Basal cell carcinoma A common kind of skin cancer.

Beringia The region where Siberia meets Alaska.

Bonobos A species of great ape related to chimpanzees.

C

Chemotherapy The use of anti-cancer drugs to destroy cancer cells.

Chromosomes The segments of DNA joined with protein in the nucleus of a cell. Humans have forty-six.

Chromothripsis A single catastrophic event within a cell resulting in the shattering and rearrangement of chromosomes.

Circulating tumour DNA Small amounts of DNA released by cancerous tumours into the blood stream.

Cluster A group of identical DNA fragments attached to a flow cell for sequencing.

Commensal A relationship between organisms where one organism gains benefit and the other organism is not affected.

Complementary single-stranded DNA (cDNA) DNA synthesized from a single-stranded RNA template.

Conjugation The transfer of plasmids between bacterial cells by direct cell-to-cell contact or by a bridge-like connection between two cells.

Consanguineous Having the same ancestry or descent; related by blood.

Coprolites Fossil faeces (poo).

Core genes Genes common to all members of the same species that perform essential basic functions for that organism.

CRISPR Clustered Regularly Interspaced Short Palindromic Repeats.

CRISPR–cas9 A technology used to edit parts of the genome by removing, adding, or altering sections of the DNA sequence.

D

DECIPHER DatabasE of genomiC varIation and Phenotype in Humans using Ensembl Resources.

Deletions A type of mutation where part of a chromosome or a sequence of DNA is lost during DNA replication.

De novo New. In genomics de novo mutations mean new mutations.

De novo sequencing Sequencing a new genome that has not been sequenced before.

Denisovans An extinct form of early human who lived in Asia from >400,000 to 50,000 years ago. Denisovans interbred with modern humans so that people in some parts of the world today carry 0.2–3 per cent of Denisovan DNA in their genome.

Differentiation A process by which a cell changes or develops into a different form to carry out a specific function.

Diploid Having two sets of chromosomes.

DTC genetic testing Direct-to-Consumer genetic testing.

E

ENCODE Encyclopaedia of DNA Elements.

Endemic In terms of infections, a disease which commonly occurs in a particular region or country.

Essential genes Genes without which an organism is unable to survive.

Exome sequencing Sequencing of just the genes or coding DNA in a genome.

Expressed In genetics, the use of a gene by making an RNA copy and then a protein from this RNA.

Expression levels Of genes—levels of gene activity resulting in a functional protein being produced.

F

Fertile Crescent A crescent-shaped region stretching from the Persian Gulf to the Mediterranean, one of the areas where agriculture originated.

Fit (fitness) The ability to survive to reproductive age, find a mate, and produce offspring.

Flatworms A phylum of worms characterized by a long, simple, flattened body which lacks blood vessels, and either a digestive tract with a single opening or no digestive tract.

Flow cell A glass slide used to sequence DNA.

Foetus A stage of development in the uterus—in humans, from the ninth week after fertilization to birth.

G

Gametocyte A sexual form of the plasmodium parasite.

Gene drive A gene-editing technique used to spread a particular trait throughout a population using biased inheritance, where the offspring within the population have more than the natural 50 per cent chance of inheriting that trait.

Genetic associations A genetic variant associated with a disease or trait.

Genetic counsellor A health-care professional who provides support, information, and advice about genetic conditions.

Genetic diversity The level of genetic variation in a population.

Genome An organism's complete set of genetic instructions. Each genome contains all of the information needed to build that organism and allow it to grow and develop.

Genome assembly The process of putting back together DNA fragments that have been sequenced, to produce a complete genome sequence.

Genomics The study of the structure and function of genomes.

Germline The genetic material being passed from generation to generation through the germ cells (sperm and egg).

Guide RNA (gRNA) A small piece of pre-designed RNA sequence used in CRISPR-Cas9 genome editing. It 'guides' the Cas9 enzyme to the desired part of the genome.

H

Haplotype A group of genes, located on one chromosome, inherited together from a single parent.

Helminths Multicellular, worm-like, free-living or parasitic organisms.

Heterozygous An individual who carries two different alleles for a particular gene.

Hierarchies Systems where people are ranked according to their perceived status.

Homo sapiens The binomial name for modern humans.

Homologous Descended from a common ancestor; being similar.

Homologous recombination A process in which nucleotide sequences are exchanged between two similar or identical molecules of DNA.

Homology-directed repair When a cell uses a very similar piece of DNA to help repair a break in a DNA strand.

Homozygous An individual who carries two of the same alleles for a certain gene.

Horizontal acquisition A process in bacteria where DNA is acquired by one bacterium from another bacterium that is not the direct parent of the cell.

Human Phenotype Ontology Terms (HPO terms) A set of standardized terms used to describe phenotypes or characteristics seen in human disease.

I

In silico Performed on a computer or via computer simulation.

Integration A process where bacterial cells use recombinaze enzymes to incorporate new sections of DNA into the genome.

K

Kb Kilobase—meaning a thousand base-pairs.

Kinetoplastids A group of flagellated, single-celled protists, including some that are parasites, characterized by a structure called a kinetoplast (a large mitochondria with many copies of the mitochondrial genome).

Knockout gene A technique in which one of an organism's genes is changed or 'knocked out' so it no longer functions.

L

Lactose The sugar in milk.

Linkage analysis A process that looks for parts of the genome that always segregate with the disease phenotype in a family of affected individuals.

Longitudinal studies Research involving information about an individual or group gathered over a long period of time.

Lymphoma Cancer of the lymphatic system, causing lymphocytes to change and grow out of control.

M

Mb Megabase—meaning a million base-pairs.

Melanoma A type of skin cancer.

Metabolic reconstruction A process where all the genes that code for enzymes are mapped to show all the metabolism pathways and protein interactions in a cell.

Metagenomics The study of all genomes within a community, such as a microbiota.

Microbiota A community of microorganisms found in a particular environment, such as the human gut.

Microsatellite markers Sequences of repetitive DNA found within the genome.

Molecular clock A term for a technique that uses the mutation rate of DNA sequences to calculate points in time when species diverged.

Monogenic Involving or controlled by a single gene, for example monogenic diseases such as Huntingdon's disease are caused by changes to a single gene.

Mosaicism Mutations that are present in some cells in an individual but not in others, for example in germ cells (sperm or eggs) but not in other tissues. This can lead to the presence of two or more populations of cells with different genotypes in one individual who has developed from a single fertilized egg.

Multigene families Genes that exist in multiple copies in the genome, with the copies being very similar (but not identical) to one another.

Mutation Any change to the DNA sequence.

Mutation rate The number of changes to a piece of DNA (such as a genome), measured in generations or years.

N

Natural selection The process by which the organisms best adapted for an environment survive and pass on their genes to the next generation. This is the mechanism of evolution.

Neanderthal An extinct form of early human who lived in Europe and parts of Asia from >400,000 to 40,000 years ago. Neanderthals interbred with modern humans so that many people today carry 2 per cent of Neanderthal DNA in their genome.

Neolithic Literally referring to a period when a new type of stone tool was used, the term is more generally used to refer to the adoption of food production by societies. The period began around 10,000 years ago in several parts of the world.

Neutral mutation Changes in DNA sequence that are neither beneficial nor detrimental to the ability of an organism to survive and reproduce.

Next-generation sequencing (NGS) A term used to describe a number of different post-Sanger capillary-sequencing technologies.

Non-synonymous mutation A change in the DNA sequence that causes a change in the encoded amino acid in a protein sequence.

Nucleotide content The proportion of particular nucleotides (adenine and thymine, guanine and cytosine) in a genome.

Nucleotide excision repair A process where the cell repairs damaged DNA.

Null model A pattern-generating model that is based on randomization of data.

O

Opposable Refers to our thumbs, which we humans (and some other apes) can move independently to touch our fingers and thus carry out precision tasks.

P

Palaeontologists People who study fossils.

Pan-genome The total set of genes in all strains within a bacterial species.

Paratenic hosts An intermediate host for a parasite in which no development occurs, but which is still needed to maintain the parasite life cycle.

Pathogenicity islands Clusters of genes within a bacterial genome that are mobile and have a specific function.

Phase variation A process that helps bacteria to quickly generate diversity and adapt to rapidly changing environments by randomly switching on and off proteins.

Phenotypes The physical traits and characteristics of an organism resulting from their genetic makeup.

Plasmids Small, circular, independently replicating pieces of DNA found in bacteria.

Population bottleneck An event that drastically reduces the size of a population.

Positive natural selection The selection of genetic variants that result in advantageous phenotypes.

Protista The kingdom comprised of mostly singled-celled eukaryotes, such as the Protozoa.

Protozoans Single-celled eukaryotes, either free-living or parasitic.

Q

Quantitative Trait Locus (QTL) A region of DNA which is linked to a particular characteristic in the phenotype.

R

Radiotherapy A cancer treatment where high-energy rays are used to damage cancer cells and stop them from growing and dividing.

Recognition sequences In the context of gene editing, a specific sequence of DNA that a nuclease enzyme will recognize and cut.

Recombination A process where pieces of DNA are broken and rejoined to produce new combinations of alleles.

Retinoblastoma A rare type of eye cancer that can affect young children.

Reverse vaccinology Using bioinformatics tools to screen an entire pathogen genome to identify potential vaccine targets.

Rickets A disorder of the bones caused by insufficient vitamin D, leading (in severe cases) to deformities.

RNA interference (RNAi) A biological process in which RNA molecules prevent gene expression or translation taking place.

Roundworm A small, free-living or parasitic worm characterized by a long, round, unsegmented body, a mouth, and digestive tract leading to an anus.

S

Scavenge In animals, using food discarded by other species, such as the bones of animals killed by predators.

Selection pressure External factors that affect an organism's ability to survive in a particular environment.

Sexual selection When a feature is selected for because it increases the chances of an individual finding a mate.

Sickle cell allele The recessive allele which codes for sickle-shaped red blood cells.

Single nucleotide variation (SNV) A single letter change or mutation in a DNA sequence, such as switching a C for a T.

Synonymous mutation A change in the DNA sequence that does not result in a change in the encoded amino acid in the protein sequence.

T

Transcription The first step during protein synthesis when the DNA in a gene is copied to produce an RNA transcript called messenger RNA (mRNA).

Transduction A process by which a phage virus transfers DNA from one bacterium to another.

Transformation A process where bacteria take up DNA from their environment.

Translation The second step during protein synthesis where the mRNA molecules are used as guides to make proteins.

Translocation Movement of a section of DNA from one place in the genome to another.

Trios Sequencing structure to diagnose possible causes of a genetic condition—a trio is made of both parents and an affected child.

Trisomies Chromosomal disorders characterized by an additional chromosome, so three copies of a chromosome are present. Down syndrome is the result of trisomy.

Tumour suppressors A type of gene that plays a role in protecting cells from cancer.

V

Variant A letter or region of a gene that is different in different individuals.

Variant-calling algorithm An algorithm used to identify **variants** from sequence data.

W

Whole-exome sequencing (WES) Targeted sequencing of just the genes or coding DNA in a genome.

Whole-genome sequencing (WGS) Sequencing all of the DNA in an organism's genome.

INDEX

1,000 Genomes Project 8
100,000 Genomes Project 51, 57

acrylamide gel 153
adaptive immune responses 104
adaptive introgression 145
additional looked for findings 51
adenine 2, 32, 150
agriculture 142–3
allele frequencies 14
alleles 7, 108
altitude sickness 145
amino acids 32, 133
amplifications 29
ancient DNA (aDNA) 79, 123, 131–2,
    137, 138, 142
Ancylostoma duodenale 119
animal modelling 9, 20–2, 119
    alternatives to laboratory
        animals 119–20
Anopheles mosquito 103, 108
anthelmintics 119
antibiotic resistance 67, 75, 79, 88–9
antibiotics 86–9
antigenic variation 103–7
ape species 124–5, 126
apicomplexa 99
apoptosis 30, 36, 36
Apple watch 57
archaeological specimens 79, 125–9
aristocholic acid 32
array-cgh 18
array techniques 8, 9, 11
artificial DNA sequences 168–9
Ashkenazi Jewish community 7
Australia 49
Australopithecus afarensis 128
autism 135
autoradiograms 153–4

Bacillus anthracis 75
bacterial genomes 65, 72, 73, 75–81
        (*see also* pathogen genomics)
    metagenomics 83–5
    programmed variation 81–3
balancing selection 144
barbers pole worm (*Haemonchus
    contortus*) 101, 118
Barth syndrome 9
basal cell carcinoma 35, 39
BCL2 gene 30

BCL2 translocation 38, 41
BCR-ABL fusion genes 41
BCRABL protein 30
Beringia 135
'big data' 58
bilharzia 113
blood flukes 113–18, 119
bloodstream expression sites
        (BESs) 106–7
bonobos 124, 126, 127
Bordetella bronchiseptica 78
Bordetella pertussis 78
BRAF gene 41
BRAF pathway 39, 40
brain size 134–5
BRCA1/BRCA2 genes 12, 33, 36, 49,
    54, 56
breast cancer 12, 33, 36, 41, 42, 49,
    50, 56

CADD 12
cancer
    breast cancer 12, 33, 36, 41, 42,
        49, 50, 56
    carcinogens 32
    childhood cancers 41
    circulating tumour DNA 42
    defences against 35–7
    development 33–5
    dIagnosis 41, 42
    DNA analysis 4, 25
    drug discovery 39–40
    Ewing's sarcoma 41
    family cancers 33
    genetic origins 27
        amplifications and whole
            genome duplication 29
        deletions 29
        genome shattering 31
        mutations and signatures 31–3
        single nucleotide variations
            (SNVs) 28–9
        structural variants 162
        translocations 29–30
    leukaemia 30, 41
    lung cancer 27, 32
    lymphoma 30, 41
    phenotype of cancer cells 38
    risk 51
    skin cancers 31, 32, 33–5, 35–6, 40
    stages 25, 26

symptoms 25
    treatment 38–9, 41–4
        discovery of new
            treatments 44–5
        failure 43–4
        viruses 31, 32
capillary sequencing 156–7
carcinogens 32
Cas9 enzyme 166, 167
CDK13 gene 8–10
Chad 102
Chalkhill Blue butterfly 139, 141
chemotherapy 38
childhood cancers 41
chimpanzees 124, 125, 126, 127,
    132, 133, 134
Chinese medicines 32
cholera 79
chromosomes 124, 150
chromothripsis 31
chronic myeloid leukaemia 30
circulating tumour DNA 42
climate change 142
Clostridium difficile 84, 89
clusters 159
collaboration 12–13
colonization resistance 84
commensal bacteria 82
commercialization 57–8, 62
complementary single-stranded DNA
        (cDNA) 164
conjugation 76, 77
consanguineous populations 7
consumer decisions 55–7, 62
coprolites 131
counselling 54–5, 56
Crick, Francis 3
criminal investigations 59–61
CRISPR-Cas9 45, 62, 167, 168
CrisprQTL mapping 12
cyclobutane pyrimidine dimers
        (CPD) 35
cytosine 2, 31, 150

data explosion 49, 51
data security 59, 62
data sharing 13–17, 58 (*see also*
    genomic data)
databases 13, 48, 49, 59
Davies, Dame Sally 4, 47
dbSNP database 13